Structure-Oriented Evaluation

Uranchimeg Tudevdagva

Structure-Oriented Evaluation

An Evaluation Approach for Complex Processes and Systems

 Springer

Uranchimeg Tudevdagva
Chemnitz University of Technology
Chemnitz, Germany

Mongolian University of Science
and Technology
Ulaanbaatar, Mongolia

ISBN 978-3-030-44808-0   ISBN 978-3-030-44806-6   (eBook)
https://doi.org/10.1007/978-3-030-44806-6

This Springer imprint is published by the registered company Springer Nature Switzerland AG.
The registered company address is: Gewerbestrasse 11, 6330 Cham, Switzerland

Foreword

In our modern and industrialized world, the quality of products is most important. Testing methods have been explored for a long time and a huge variety of test methods can be found in literature and praxis. The idea is to ensure the quality and reliability of products when they have been produced. Two main drivers of these approaches can be identified. First, the minimization of "Time to Market" requires a stable production process and low failure rates. Second, high-quality products will be sold at higher volumes, this leads to higher profit and bigger markets. Over many different kinds of products, this concept has been found valuable.

In the last decade, more and more software products can be found which are the implementation of processes. Of course, the functionality of process implementation must be ensured. This includes process steps, data consistency, and process control. But the quality of a process implementation cannot be explored by the traditional quality and testing methods. The main reason can be found in the various interest groups which are involved as the process owner, process user, or in any other role.

Once education technologies came into the focus of digitalization and optimization, we see that no quality process is available. Learning based on education technologies can simplify access to information and learning support. But learning results cannot be easily related to the usage of education technologies.

Prof. Dr. habil. Uranchimeg Tudevdagva presents in this book a well-structured method to set up a quality model for complex processes. The key idea is to understand the process structure and the different roles of the involved user groups. The new model allows to add process information to each process steps. By automated analysis methods, various parameters can be evaluated. This approach is generalized and can be applied to many kinds of processes.

In this book, an example taken from education technologies is presented as a use case. The model is set up and the processes are evaluated. The results are the basis for optimization of teaching methods based on education technologies. New research ideas put the learner into the center of the overall learning process, which is called Learner-Centered Learning (LCL). Such methods can be evaluated by the SURE module in an early development status and the relevant parameters can be optimized.

In praxis, the SURE module has been used in many international education projects. Learners (students) with different background and education status have been involved as well as teachers with varying teaching methods. It has been found that the SURE module is of great benefit for improving the learning and teaching process.

This book is highly recommended for those who deal with complex processes. Evaluation of quality parameters becomes more easy and transparent.

I am happy to recommend this interesting book to the reader and developer of complex processes.

Chemnitz, Germany   Prof. Dr. Dr. h. c. Wolfram Hardt

Acknowledgment

At the end of this foreword, I thank all who in one way or another contributed to the completion of this book. It would not have been possible to write this book without the help and support of the people around me.

I am so grateful to my family, my husband Oyunbat Tumurbaatar and my children Gantogoo Oyunbat and Oyunchimeg Oyunbat for their understanding and great patience at all times.

I wish to thank K.-H. Eger for his valuable hints during the writing of this book and fruitful discussions. I am very thankful to U. Eger for her support and patience during my writing.

Contents

List of Figures

Chapter 1
Introduction

1.1 Evaluation

Evaluation is a complex topic that increasingly touches almost all areas of life. After World War II this subject became more and more an independent discipline to which researchers from various fields contributed theoretical ideas and developed corresponding models.

Many countries have heavily invested in education and science to alleviate the consequences of war. Appropriate funds were provided for corresponding programs. The involved stakeholders and decision makers had to prove that their activities and investments have achieved their goals (Madaus et al. 1984; Stufflebeam 1994). For this purpose and as decision maker support for future investments, adapted evaluation methods were needed. Main aim of evaluations was to show value and impact of running or implemented programs.

An evaluation process covers several areas. Evaluation can be done for recognition of a program need or service and can support stakeholders to make better decisions for investments into future programs or services. This kind of evaluation is focused at need of a program. Other area for evaluation is the implementation phase of programs. Sometime this kind of evaluation is called as process evaluation. By this evaluation evaluators try to identify progress of on-going programs. A further area of evaluation is the so-called summative evaluation which is targeted at the outcome or impact of a program.

In general, one distinguishes two main directions at evaluation: formative evaluation and summative evaluation. Formative evaluation is a kind of process evaluation focused to find out how a program or process is running, does the obtained result meet the expected outcome. For example, if a given e-learning course consists of several basic modules, then all these modules should be included into evaluation, and if one of the modules did not reach its target, then the entire process has failed its target. That should also show up in the evaluation. All questions of formative evaluation are targeted to find out what supports the achievement of

© Springer Nature Switzerland AG 2020
U. Tudevdagva, *Structure-Oriented Evaluation*,
https://doi.org/10.1007/978-3-030-44806-6_1

the process goal, and how the single modules contribute to the achievement of the process goal.

Summative evaluation is also called outcome evaluation. Focus of this kind of evaluation is to find out whether the program has reached the planned goals. For example, in an e-learning course consisting of several modules targeted to increase e-communication skills of learners, the evaluator has to find out whether that course actually leads to an increase of e-communication skills among the learners.

In addition to these two evaluation directions an impact evaluation for programs can be performed. The impact evaluation is in some cases identical with summative evaluation. Nevertheless, there are some differences. The main difference is that the impact evaluation examines in more detail the outcome of the program in terms of long-term effects.

In last five decades evaluation became an important element of science and research. Many researchers worked on evaluation models and presented various types of evaluation methods and solution approaches. Depending on the application field evaluation models can be very different.

Researchers of different branches use the term of evaluation in various variants depending on their research directions and study fields. General meaning to use this term in scientific environment is to identify success of new ideas in implementations and practical applications.

Moreover, investors and stakeholders have different views to this term. As a rule they use evaluation as an instrument to control and to measure the success of their activities.

In general, evaluation is understood as a special method of quality measurement. Frequently the obtained result is used for a ranking system as quality indicator.

Experts define evaluation as follows: "Evaluation is the process of determining the merit, worth, and value of things and evaluations are the products of that process" (Scriven 1991). "Program evaluation is the use of social research procedures to systematically investigate the effectiveness of programs" (Rossi and Wright 1979).

Most important issue of evaluation is to measure merit, worth, and value of programs. Programs can be in different forms: projects, processes, and systems, for instance. For each situation an adapted evaluation method or model is needed.

Until today many researchers developed models and methods for evaluation. Therefore, author did not target to compare the here considered structure oriented evaluation model with other existing models.

The aim of this book is to describe the main concept of structure oriented evaluation model, its theoretical background, its advantages as well as some real and simulated examples for application. In contrast to the linear models, this model requires no weighting factors, so that the subjective influence of weighting factors on the evaluation result can be avoided. Moreover, a web-based tool which has been developed specially for the structure oriented evaluation model is presented in this book for the first time.

1.2   Short Review of Evaluation Models

This section covers some of the models. Selection of presented models is based on evaluation community recommendations and accessible reading and writing materials on the internet and by citation of scientific papers in the field of evaluation.

- *Objectives-oriented model.*

  Ralph W. Tyler presented objectives-oriented model in 1949 (Tyler 1949). Main focus of this model is directed to clarification of measurable objectives and data collecting based on those goals to validate expected results of program. This model was widely accepted by scholars and evaluation community and later became basis for many other evaluation models which are oriented to goal and objectives of programs. Objectives-oriented model suggests what should be done to improve educational practice. The Tyler model emphasizes consistency among objectives, learning experiences, and outcomes. Tylerian model consists of the following phases: definition of broader goals/objectives; classification of the defined goals/objectives; definition of objectives in sense of behavior; definition of situations for objective achievement; choose corresponding measurement techniques; collection of performance data and comparison of collected data with behavioral states. Advantages of this model are: taking consideration of all stakeholders, learners can participate actively, easy to understand, and defining objectives clearly.
- *Kirkpatrick's four-level model.*

  This model was developed by Donald Kirkpatrick in 1954 (Kirkpatrick and Kirkpatrick 2006, 3rd edition). Main focus of this model is to measure effectiveness of a training program. Model consists of four levels: Reaction, Learning, Behavior, and Results. Later on with minor change this model became one of standard evaluation model for programs in business and industry. In reaction level focus is on the reaction of the participants to the training or the learning experience. At the learning level focus moves to the new learning that results from the training. During behavior level the focus is on the transfer of learning and at results level the focus is on the targeted outcomes. In reaction level data is collected as feedback from learners about training program. In other words here satisfaction of participants is measured. In learning level measurement happens via performance or testing. This level has to figure out how to increase knowledge or skills of learners by training. Data collection of behavior level can occur later in three up to six months of time after training completion. Reason of this late data collection is rented to the transfer of knowledge into working place which has to be identified by evaluation of this level. Evaluation of results level focuses on specific outcomes that were targeted in training.
- *Consumer-oriented model.*

  This model was developed by Michael Scriven in the late 1960s (Scriven et al. 1967). The primary focus of consumer-oriented evaluation is around valuing a product for public. This approach is concerning to answer to the question:

how good is this product? Scriven offers to use set of criteria as the key evaluation checklist. The checklist applied as synthesizer an overall judgment of value from large amounts of data. Consumer-oriented evaluation relies on transparent and quantitative methods evaluated by an evaluator with expertise in judging things not with particular content. The consumer-oriented evaluation model is predominantly a summative evaluation approach. Next checklist is defined by Scriven: Need (Justification), Market (Disseminability), Performance (True field trials, true consumer, critical comparisons, long term, side effects, process, causation, statistical significance, educational significance), Cost and cost effectiveness, and Extended support

- *CIPP model.*

The model was developed by Daniel Stufflebeam in the 1960s together with colleagues (Stufflebeam 1968). CIPP stands for context, input, process, and product. This model supports administrators who are facing four kinds of decisions: planning decision about context, structuring decisions for selection of strategy and resolving problem, third is implementing decisions focus at implementation, planning and monitoring of program, finally product evaluation serves to guide continuous monitoring, modifying, adapting, and future thinking about the program. By CIPP model decision is more important than objectives. This model focuses on the future of a program. By this evaluation model long-term impact of the model is to be measured to make some decision about program future. Result of this evaluation should answer to questions: was the program effective and what parts of the program were effective. Evaluator in this case serves as stakeholders, policymakers, administrators, managers, board members, and other managerial staffs. Data collection techniques are quantitative and qualitative.

- *Taba's model.*

Taba's evaluation model is also known as instructional strategies model (Taba 1962). By Taba the evaluation process should care teacher's contribution to educational curriculum or program development and implementation. Based on that idea Taba offered seven major steps in evaluation. There are seven steps of model: diagnoses of learners' needs and expectations of a larger society, formulation of learning objectives, selection of learning contents, selection of learning experiences, organization of learning experiences and determination of what to evaluate and ways to do it. Main highlight of this model was to include teacher role, instructional design into evaluation process. The model includes an organization of, and relationships among five mutually interactive elements–objectives, content, learning experiences, teaching strategies, and evaluative measures (Fred 2011).

- *Discrepancy model.*

This model was developed by Malcolm Provus in 1969 (Provus 1969). The discrepancy model focus at reasons for observed actions. Aim of this evaluation model is to use discrepancy information to identify weaknesses of the program. In this model evaluation process will care about why it happened. A program examines in whole on each stage of development and those stages defined by Provus as: design, installation, process, product, and cost–benefit analysis (Boulmetis and Dutwin 2011). Five steps are needed to implement discrepancy model: S—standard; define goals of instructional system, P—performance; determine how well-identified goals are achieved, how and why, C—compare; determine gaps between what is and what should be, D—discrepancy; prioritize gaps recording to defined criteria and D—decision; determine which gaps instructional needs and which are need design and development instruction. Discrepancy model is the summative evaluation.

- *CIRO model.*

  CIRO stands for context, input, reaction, outcome. Another four-level approach has been developed by Warr, Bird, and Rackham and published in 1970 first in book Evaluation of Management Training (Warr et al. 1970). By CIRO evaluation model measurements are taken before and after training and by this it is different from Kirkpatrick's four-level model. Training objectives can have three different levels: the ultimate, intermediate, and immediate objectives. These became context for CIRO model. Input can be analysis of effectiveness of training program. The reactions of participants of training count as reaction for CIRO model. Finally, outcomes are evaluated in sense of what happened after the training.

- *Goal-free model.*

  The model was introduced by Michael Scriven in 1972 (Scriven 1973). Goal-free evaluation is the process of determining objects' merit intentionally without reference to its stated goals and objectives. The idea behind goal-free evaluation is finding out what the program is actually doing without being cued as to what it is trying to do. If the program is achieving its stated goals and objectives, then these achievements should show up, if not it is argued, they are irrelevant. It is an external evaluation model. Task of evaluator is to evaluate the impact, effect, or outcomes of program. Evaluator should be outside of program team and should have no any prior knowledge of the goals and objectives as blank slate (Boulmetis and Dutwin 2011). This is the basic requirement to apply goal-free evaluation into practice. Evaluation result shows what has happened with this program.

- *Transactional model.*

  The transaction model introduced by R.M. Rippey in 1973 (Rippey 1973). This is more of a subjective model where an evaluator plays roles of the evaluator and participants during evaluation process. By means evaluator involves into program actively and provides permanent feedback and acts as a staff member in the team. Main techniques of data collection in this model are observation and active interviews. The transactional model refers to goal-based evaluation models.

- *Eisner's connoisseurship/criticism model.*

  Elliott Eisner offered this model in 1976 (Eisner 1975). This model focused to educational programs. Main concept of model is evaluator should operate much like critics who critique art. To apply Eisner's model evaluator must become connoisseurs—that is, specialists or experts—capable of seeing nuances in the programs they study, nuances that ordinary viewers are likely to miss (Donmoyer

2014). Connoisseurship is private, but criticism is public noted Eisner (1985). By his idea critic should help to improve program.

• *Logic model.*

Early 1960s started discussions on logic model in evaluation literature (Suchman 1967; Wholey 1979). Later in 1998 W.K. Kellogg Foundation published Logic Model Development Guide: Using Logic Models to Bring Together Planning, Evaluation, and Action (Kellog 2004). The logic model visualize in systematic way relationship among the resources, activities, and changes of program. The logic model helps to break down complexity of program. It helps to understand what type of measurement to look at, what can be evaluated, what can be realistically measured. This is a consultative and collaborative process. When any program starts stakeholders expect to see some change after program. Designing a logic model is an excellent way to simplify the complexity of program and assimilate the causal linkages that are assumed to occur from the start of the program through to the impact it makes.

• *Five level ROI model.*

A well established and wide applied evaluation model: the four-level Kirkpatrick's model became basis for five level ROI model by Jack Phillips in 1980 (Phillips 1991). The new contribution from Philips was the addition of a fifth level for calculation of investments into training program. Aim of this new level was to show value of training program and to do cost–benefits analysis. ROI stands for return on investment. By this idea five level ROI model differs from Kirkpatrick and Kaufman's five level models. ROI model focuses at data collection, isolation of the impact, accountable benefits, and calculation of return on investment (Deller 2019a).

• *Decision-making model.*

This model has been developed by Daniel Stufflebeam in 1983 together with Madaus and Scriven (Stufflebeam 1983). Main aim of decision-making model is to make a better decision regarding the future of program. By this evaluation model how the program is actually running is not on the focus. Advantage of decision-making model is opportunity to use different kind of techniques for data collection. The quantitative methods like tests and records can be applied for this model. Alternatively, qualitative methods such as interviews, observation, and surveys can be used for this model.

• *Kaufman's five levels model.*

This model was developed based on Kirkpatrick's four-level evaluation model. Roger Kaufman published the paper "Levels of evaluation: Beyond Kirkpatrick" in 1994 and it was an introduction to Kaufman's five levels evaluation model. Kaufman had an idea to divide the first level of Kirkpatrick's four-level evaluation model: Reaction level divided into two components: Input and Process. Moreover he added a fifth level for evaluation (Kaufman et al. 2007). By definition of Kaufman Level 1a stands for Input, Level 1b stands for Process. Level 2 stands for Acquisition, and Level 3 is for Application. Level 4 is Organizational payoffs and last Level 5 stands for Societal Outcomes (Deller 2019b). In some literature noted that Level 2 and 3 are Micro levels, Level 4 is a Macro level, and Level 5 is a Mega Level.

- *People-process-product continuum model.*

The e-learning P3 model developed by Badrul H. Khan in 2004 (Khan 2004). By his idea e-learning is process where involved people are responsible for products, therefore for evaluation of e-learning has to cover all these three aspects. People play key role in this model. That can be: content experts, instructional designers, project managers, and programmers. The e-learning process includes analysis of target groups, learning materials, planning and designing, development and deployment of course. All these stages should include internal evaluation. E-learning product always needs improvement. In that sense evaluation of e-learning should be a continuous process. Based on results of formative evaluation learning materials should be revised. E-learning P3 continuum model can be comprehensive evaluation map e-learning programs.

- *The unfolding model.*

Valerie Ruhe and Bruno D. Zumbo developed in 2008 the unfolding model (Ruhe and Zumbo 2008). This is an evaluation model for e-learning and distance courses. The unfolding model includes four components: scientific evidence, cost benefits, underlying value, and intended outcomes. The model can be adapted to different course designs and technologies. Evaluators can pick and choose from among the tools presented to tailor the evaluation to their own needs. This feature is important as the technology evolves permanently and accordingly to that an evaluation model for e-learning should be able adaptive.

- *PDPP Model.*

The PDPP stands for planning, development, process, and product (Zhang and Cheng 2012). The PDPP model is an evaluation model for e-learning. The model was introduced by Weiyuan Zhang and Y.L. Cheng in 2012. Basis for this model was CIPP model. By their definition the e-learning evaluation PDPP model consists of four stages which include 26 components. There are: the planning stage with market demand, feasibility, target student group, course objectives, and finance and quality assurance. The development evaluation stage includes: blueprint, e-learning platform, course website, instructional design, learning resources, assignment and examination and tutors. The process evaluation consists of overall evaluation, technical support, website utilization, learning interaction, resources utilization, learning evaluation, learning support, and flexibility. The product evaluation includes satisfaction degree, teaching effectiveness, learning effectiveness, other outcomes, and sustainability.

- *Expertise-oriented model.*

The expertise-oriented evaluation model is one of the oldest models in evaluation (Fitzpatrick and Sanders 2011). The evaluation is based on feedback provided by experts. Experts are people who are recognized as having knowledge, experience, and authority within their fields. Experts for evaluation may be chosen to be part of a panel of experts who form an evaluation committee. Nowadays, accreditation is the most common formal professional evaluation in the world. In the accreditation process many professions are included to ensure quality in their fields. Any organization prepare documents as internal evaluation report for accreditation and later experts visit educational institution to approve the report and judge the accuracy of the self-evaluation results.

- *Participant-oriented model.*

  The participant-oriented evaluation is an evaluation approach focused primarily on serving the needs, interests, and values of those participating in the program (Fitzpatrick et al. 2004). During this evaluation approach stakeholders are directly involved in the evaluation and information is gathered directly from participants. Characteristics of this approach are dependent on needs, values, and questions of stakeholders. It covers a wide range of sources and is adaptable to many situations. The model helps stakeholders to improve their skills and understanding of evaluations.

- *Theory-driven model.*

  The theory-based evaluation pioneers are Glaser, B.G., Strauss, and Weiss (Glaser and Strauss 1967; Weiss 1998). The theory-driven (based) evaluation denotes an evaluation strategy or approach that explicitly integrates and uses stakeholder, in social science some combination of other types of theories in conceptualizing, designing, conducting, interpreting, and applying an evaluation (Coryn et al. 2011). The aim of theory-based evaluation is to gain a better understanding of the program, to present details of reasons, to clarify the relationship between the program problems and the program's actions (Fitzpatrick et al. 2011). Theory-based evaluation addresses the input and the output of evaluation factors as well as all other factors in between. Stewart Donaldson defined seven steps for theory-based evaluation process (Donaldson 2012). These are: engage relevant stakeholders, develop first draft of program theory, present draft to stakeholders, plausibility check, communicate and revise findings, probe from model specificity, and finalize program impact theory.

1.3 Structure Oriented Evaluation Approach

Main concept of structure oriented evaluation model (SURE model) is based on a multidimensional understanding of evaluation process and a corresponding analysis (Tudevdagva and Hardt 2011). First two steps of the SURE model is definition of evaluation goals. A complexity of definition of evaluation goals request some transparent method to discuss evaluation goals (Tudevdagva and Hardt 2012). To fulfill this requirement came out the idea to use logical schemes as structure for evaluation goals in the SURE model (Tudevdagva et al. 2012).

Evaluation of e-learning should consider views of all involved groups. This is not a completely new view. Many evaluation models offer to take care the interests and views of involved groups as well as evaluators. These have different names in respective models, for instance: teacher, consumer, people, participants, stakeholders, and society (see Tyler, Scriven, Stufflebeam, Provus, Taba, Kirkpatrick, Kaufman, and others). Hereinafter they are called as involved groups.

By SURE model all these players are called as involved groups. They are different groups who are connected to the evaluation process via their interests. The SURE model belongs to goal-oriented evaluation models. Main focus of SURE model is to define corresponding key and sub goals of evaluation and to organize these goals in sense of the main goal of evaluation. Key goals must be achieved so that the main goal of evaluation process is reached. In this sense they are connected as a logical series structure together. A sub goal is a goal which is assigned to a key goal where the key goal is reached if at least one of the assigned sub goals is successfully reached. Hence the sub goals of a key goal are connected vial a logical parallel structure together. By means of key and sub goals very complex goal structures can be described. The SURE model is now a way of assessing how such a process has achieved its goal on the basis of observation data, taking into account the respective logical structure of the process. Main characteristic of SURE model is that it has an own data processing theory which is adapted to goal structure of process.

Logical structure of evaluation goal or process helps to involved groups figure out their view and expectations for evaluation processes correctly.

Some evaluation models are directed to selected target groups like stakeholders, consumers, or participants: CIPP model, Taba's model, P3 model, expertise-oriented model, participants' oriented model, for instance. The SURE model is not designed for specific, selected target groups. Opposite, the SURE model tries to cover expectations and view of all involved groups in the evaluation process.

To describe the main concept of the SURE model in more detail, it is considered an abstract example. For this purpose, it is assumed that an e-learning course is to be evaluated.

Preparing and running an e-learning course is a complex and time-consuming process that in many cases requires additional investment. After completion of the content and multimedia design, the course must be embedded into a corresponding e-learning platform. This happens within the framework of a suitable learning management system which includes all necessary management elements, from content design through multimedia implementation to enrollment in the course. An effective learning management system is also important for constructive communication between tutors and learners.

A successful implementation of an e-learning course is not possible without contributions of the following groups:

- professors or tutors, who are owner of course contents;
- multimedia group which transfers the provided learning materials into a digital version;
- decision makers who provide the investments;
- educational institutions which offer the course by their learning management system and learners who takes part on the e-learning course.

Main task of evaluator is to consider all views of groups involved in evaluation process. Each group has different views and expectations regarding the implementation of e-learning course. But all groups have a common key target. They all are interested in a successful evaluation of their course. Initially thereby each group has its own ideas and criteria for evaluation. Main task of evaluator is it then to identify all ideas, expectations, and evaluation criteria of involved groups and unify those into goals of evaluation. That means all involved groups have to answer the question: What you want to evaluate?

This is a very challenging task for evaluator. All involved groups can have completely different views and understanding about evaluation targets. The reason is easy to understand. The groups involved are very different. Therefore, before planning and designing the evaluation process, the evaluator should find answers to questions from the respective groups.

Any e-learning course evaluation should collect data by unified checklist. Participation of involved groups at design of evaluation process improves acceptance of evaluation results by involved groups. In other words, if some expectations or criteria are ignored which an involved group wants to include in evaluation, that group may later reject findings of evaluation process. Successful evaluation can only be possible if the evaluation findings are useful for next activities of the groups involved. Therefore identification and consideration of all expectations and views of the groups involved as transparent as possible is a very important task for evaluator.

Views and expectations can be completely different in each group. For example, stakeholders want to know how satisfied learners are with e-learning course, professors want to know if the e-learning course is as successful as classroom teaching, multimedia developers want to know which kind of media or which digital elements provide added value for e-learning, and so on.

For example, suppose five groups are involved in the implementation of the e-learning course, and Fig. 1.1a illustrates their perspectives and expectations regarding the assessment of the course.

In Fig. 1.1 the groups are illustrated by different colors:

- Yellow group represents teachers and professors;
- Blue group represents multimedia developers;

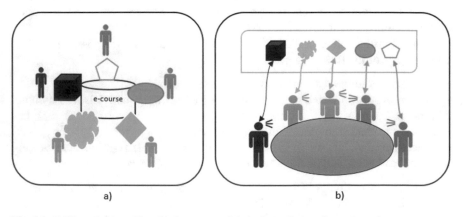

a)   b)

Fig. 1.1 Different views of involved groups and their discussion on the same topic

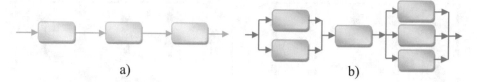

a)   b)

Fig. 1.2 Logical structures of SURE model

- Green group represents investors;
- Gray group represents educational institution;
- Black group represents learners.

There are several techniques to gather information about view and expectations from involved groups. For example, discussions, conversations, interviews, checklists, questionnaires can be used. An effective discussion form can be team meetings (Fig. 1.1b). But verbal communication is not always an appropriate channel to reach agreement. It may be that even after long discussions about a specific topic, there are still different opinions about the same project.

The SURE model uses logical structures to define views of involved groups. There are two basic structures: series and parallel (Fig. 1.2).

Such a visual view helps the involved groups as well as the evaluator to find answers to questions: What do you want to reach by evaluation? What is your goal of evaluation? What do you want to figure out by evaluation?

Behind the use of logical structures is the following idea. As a rule, a process goal can be broken down into several key goals. These are partial goals that must be reached during the process so that the process can achieve its goal. If one of these goals is not achieved, then the process as a whole has not reached its goal. Formally, this fact can be described by a logical series structure whose elements are connected by a logical AND. Because of the dominant importance of key goals in achieving a process goal, these goals must be very carefully defined.

In many cases, there are various ways or means of achieving a key goal, or various options can contribute to achieving a key goal. In this way, several sub goals can be assigned to a key goal, where the key goal is achieved if at least one of the associated sub goal achieves its goal. This fact can be described by means of a logical parallel structure whose elements are connected by a logical OR.

This logical background of the SURE model is useful for communication between the involved groups and the evaluator with regard to the formulation of the evaluation goal.

By breaking down the main goal into key and sub goals, a non-linear view of the evaluation process is achieved whereby the frequently very different interests of the groups involved can now be assigned to sub processes or structures of the overall process. In this way, for example, long and difficult discussions can be avoided, as that is usually the case when choosing suitable weighting factors for linear models.

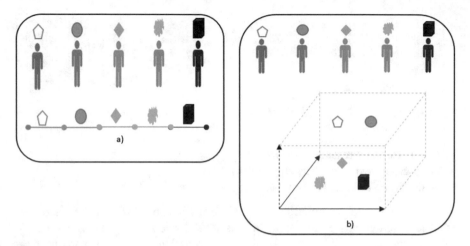

Fig. 1.3 Spaces for evaluation processes

The inclusion of the logical structure of the process and the resulting multi-dimensional approach facilitate the identification of the groups involved with the aim of the evaluation. In addition, by avoiding weighting factors and use of data processing adapted to structure of process an objectification of evaluation process takes place. Figure 1.3 symbolically illustrates the difference between linear and non-linear approaches.

The SURE model is a very flexible evaluation model and can also be used in the evaluation of complex systems. Further examples are the evaluation of robustness of embedded systems (Heller 2013) and evaluation of faculty accreditation in university (Bayar-Erdene 2019).

At application of SURE model, it is important to define the evaluation goal. Before the start of the evaluation, it must be clearly defined **what is the goal of evaluation, which structure does** the process to be evaluated have, and how this structure can be described by **key** and **sub** goals. Once logical structure of evaluation goal is clearly defined and is accepted by all involved groups the next steps are easier because of data processing rules of model are adapted to structure of process (Tudevdagva et al. 2013a,b).

To support application of SURE model an online tool has been developed by the author. This tool includes a simulator for generation of structured process data, by which the properties of SURE model can be studied (Sect. 1.4).

1.4   Evaluation Steps of the SURE Model

The application of SURE model includes eight steps (Tudevdagva et al. 2014a,b,c,d). These are to be processed in the subsequent specified order.

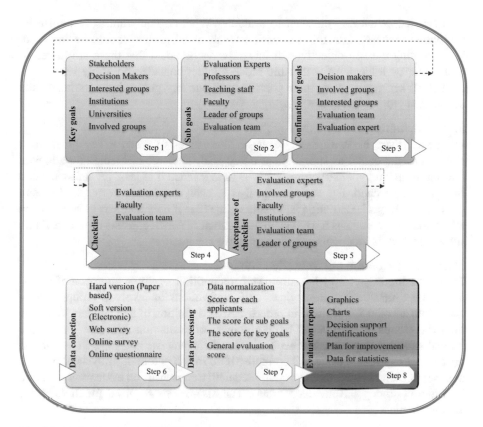

Fig. 1.4 Evaluation steps of SURE model

All eight steps of the SURE model linked whereby in logical sense together, output of the previous step becomes as input for the next step (Fig. 1.4).

These are the eight steps of the SURE model:

- Definition of key goals
- Definition of sub goals
- Confirmation of evaluation goals
- Creation of adapted checklist
- Confirmation of checklist
- Data collection
- Data processing
- Evaluation report

Below the steps are described in detail.

Step 1. Definition of Key Goals

Each evaluation starts with definition of key goals. For that evaluator has to figure out all essential goals of evaluation process. Thereby the following questions should be answered: What do you want to evaluate? What do you want to recognize by evaluation process? Into this discussion all involved groups should be included. When setting the key objectives, it must always be noted that these are linked to each other by logical AND that the process under consideration only reaches its goal if all key goals reach their goal. If only one key goal misses its target, the considered process has missed its target.

Step 2. Definition of Sub Goals
Frequently there exist different ways or alternatives to reach a key goal. Such alternatives are called as sub goals of a key goal. They are connected by a logical OR and form a parallel structure. A key goal is reached if at least one of associated sub goals is reached. For each key goal is to analyze whether corresponding sub goals exist. A special situation is, of course, the case if only one way exists to reach a key goal. Then one has a trivial sub goal structure which consists of only a single sub goal which is identical with the respective key goal.

Step 3. Confirmation of Evaluation Goals
That is an important step which follows from multidimensional approach of SURE model by which the different views and expectations of involved groups can be taken into account for evaluation process. After definition of sub and key goals the involved groups have to check the correctness of evaluation structure. Beside the logical structure they have to check whether their interests and roles in evaluation process are captured clearly and correctly. If no match is found, analysis of process in sense of Steps 1 and 2 must be continued until a consensus on evaluation structure is achieved among the involved group. The discussions required for this are not always simple.

Step 4. Creation of an Adapted Checklist
After logical structure of evaluation goals has been defined and confirmed an adapted checklist for data collection must be created. To create checklists or questionnaires there are different free and commercial tools. Important here is that the checklist for data collection is adapted and exactly fits to the defined and confirmed goal structure.

Step 5. Confirmation of Checklist
That is a further important step like confirmation of evaluation goals. This step can also be difficult, because it is sometimes not easy to find or formulate an adequate question for evaluation of a defined key or sub goal whose answer shows how successfully this goal was achieved. The questions of checklist should be formulated in discussion with involved groups and finally are confirmed by all involved groups. Sometime a certain correction of evaluation structure may become necessary.

Step 6. Data Collection
The data must be recorded in accordance with the agreed checklist. The form of data collection, online or paper-based, for example, does not play any role in this.

Step 7. Data Processing
This step is original for the SURE model. The analysis of the checklist data and the calculation of corresponding empirical evaluation scores is done according to the methodology developed in Sect. 1.2, taking into account the agreed logical structure of the key and sub goals. The corresponding calculation rules for the calculation of the empirical scores are provided there. Examples can be found in Sect. 1.3.

Step 8. Evaluation Report
This is the final step of the SURE model. After data processing, the evaluator receives several empirical evaluation scores. The most important parameter is the empirical evaluation score for the entire process with values in the interval [0,1], which mirrors an average score level of the survey participants. Beside the empirical evaluation score for evaluating the overall process, the SURE model moreover provides empirical scores for all key and sub goals as well as for each single checklist data record. Based on these values, the groups involved can see how they contribute to the achievement of process goal via the key or sub goals. On this way, the weak points and well running substructures of the process to be evaluated become clear. On the basis of all of these evaluation values, a detailed evaluation analysis can be carried out and summarized in an evaluation report.

1.5 Summary

In recent years, the demand for evaluation methods has increased steadily. This is particularly evident in the area of education and in the implementation of social programs. Parallel to this development, evaluation models have been developed under various aspects over the past 10 years.

In the first section of this chapter, some of the current evaluation models are briefly presented, taking into account internet availability, accessible literature resources such as e-books, online archives as well as articles and conference reports. The models are flown out in chronological order.

Ralph W. Tyler developed the core idea for object-/goal-oriented models. Already in the 40s he developed the idea of goal-oriented evaluation. His model was later the starting point for many other models such as Taba's model as well the participant-oriented models. The Tyler model looks at the objectives of a program, where the evaluation is aimed to find out whether the program could achieve its predefined objectives.

Donald Kirkpatrick is a further pioneer of evaluation theory who developed a basic concept for evaluation of training programs. He defined four main levels of training evaluation: Reaction, Learning, Behavior, and Results. This model is widely accepted by other researchers in evaluation field and became a kind of standard evaluation format for many training and evaluation institutions. His four-level evaluation model became basis for other famous models like Kaufman's five level model and the five level RIO model, for instance.

Michael Scriven is one of the most important researchers in evaluation theory. In the 1960s he developed a consumer-oriented assessment model. The consumer-oriented assessment model is aimed at public projects and is a transparent and quantitative method for evaluators in special application situations. Later, in the 1970s, he introduced his famous goal-free model. Scriven's goal-free model was a completely new view of the evaluation process. Earlier models usually focused on goals, objectives, and outcomes, but the goal-free model offered a different kind of assessment. In the sense of this model, evaluators should avoid obtaining information about the objective of the program.

Daniel Stufflebeam is another famous scientist in the evaluation community. He developed a complex evaluation model called the CIPP model. CIPP is the abbreviation for context, input, process, and product. This model became the development background for models such as the people-process-product continuum model; the planning, development, process, product model; and the context, input, reaction, outcome model, for example. Characteristic of the CIPP model is that it focuses on the decision-making process and not on the objectives of a program. Later, in 1983, Stufflebeam introduced another evaluation model: the decision-making model. This was the result of teamwork with Madaus and Scriven.

Later, in the 1970s, Provus developed a discrepancy model that focused on finding out the reasons for what had happened. The aim of this model was to identify the weaknesses of a program and to contribute to its improvement. Focus of this evaluation model was process and product.

A further model of this time was the CIRO model by Warr, Bird, and Rackham. Idea for this model overlaps with the Kirkpatrick's four-level model and the CIPP model in some cases. At CIRO model two measurements are performed—one before and one after training.

E. Eisner developed an evaluation model which focused not on goals and objectives of program. He recommends to use criticism for evaluation. According to his idea evaluator should be expert in that field who is able to find out the nuances in the program. Eisner's model aimed to find weaknesses in program in order to improve the program.

In the 90s, a discussion began about the use of logic models in program evaluation (Suchman and Wholey).

In 1998, the W.K. Kellog Foundation published a guide to the use of logic models in evaluation. An important aspect in the use of the logic model is the history of the program. All steps and phases of the program should have a logical connection and must be represented by logic diagrams.

One of the best-known and most widely used models for evaluating training programs is the five-step ROI model. This assessment model was developed on the basis of Kirkpatrick's four-level model. The main advantage of this model is the additional fifth level at which the return on investment of the training program is calculated. The calculation rules of the model are clear and easy to apply.

Since the 2000s, the demand for evaluation models for e-learning has grown strongly. Accordingly, this area has moved into the focus of evaluation theory. One of the first models in this direction was the P3-model of B. Khan (2004);

he presented his people-process-product continuum model. The basic idea for this model is to involve the role of the people involved and their responsibility for product quality more closely. E-learning is a process that requires constant adaptation and improvement, which B. Khan tries to achieve with the help of his continuum model.

V. Ruhe and B. Zumbo developed a further evaluation model for e-learning. Their unfolding model consists of four basic elements: scientific evidence, cost benefits, value, and results. In 2012, they introduced moreover another assessment model for e-learning, the PDPP model. The planning, development, process, and product model is similar to the P3 model. The highlight of this model is, its structure includes 26 evaluation components.

Second subsection of the first main chapter describes structure oriented evaluation approach in general. The structure oriented evaluation model started with idea to develop evaluation model where data processing should be notably focused more so than other previous existing evaluation models. Keyword "Structure" comes from idea to design evaluation goals via logical structures. Two basic logical structure defined by SURE model: Series and Parallel. Visualization of evaluation goals via logical structure improves transparent attribute of evaluation model. Logical structures are easy to understand and if a change occurs in definition of evaluation goals, it can be very visible for all involved groups of evaluation, which is not so easy to handle during other evaluation models.

Using logical structures the evaluator is trying to do an evaluation in multidimensional understanding. By idea of structure oriented evaluation model evaluator has to care all expectations of involved groups during the evaluation process. For this logical structure can support evaluator its clearness and understandable view.

Third subsection of the first main chapter explains evaluation steps of the SURE model shortly. The SURE model consists of eight steps. Each step plays important role in the evaluation process. Therefore, evaluator should follow those steps to use SURE model for evaluation. First two steps are aimed to define key and sub goals of the evaluation process. Third and fifth steps are confirmation steps for evaluation goal structure and evaluation checklist. The fourth step focuses to design adapted checklist or questionnaire. In the sixth step data has to be collected. And in the seventh step the SURE model has to compute evaluation scores. Final eighth step is for evaluation report.

References

Bayar-Erdene, L. (2019). *Evaluation of faculty members by structure oriented evaluation.* Doctoral Thesis, Mongolian University of Science and Technology, Ulaanbaatar.

Boulmetis, J., & Dutwin, P. (2011). *The ABCs of evaluation: Timeless techniques for program and project managers* (3rd ed.). San Francisco: Jossey-Bass.

Coryn, C. L. S., Noakes, L. A., Westine, C. D., & Schröter, D. C. (2011). A systematic review of theory-driven evaluation practice from 1990 to 2009. *American Journal of Evaluation, 32*(2), 199–226. Available from https://doi.org/10.1177/1098214010389321

Deller, J. (2019a). *The Complete Philips ROI Model Tutorial for Beginners*. Available from https://kodosurvey.com/blog/complete-philips-roi-model-tutorial-beginners

Deller, J. (2019b). *Kaufman's Model of Learning Evaluation: Key Concepts and Tutorial*. Available from https://kodosurvey.com/blog/kaufmans-model-learning-evaluation-key-concepts-and-tutorial

Donaldson, S. I. (2012). Evaluation theory and practice. *Connections: European Evaluation Society, 2*, 8–12.

Donmoyer, R. (2014). *Elliot Eisner's lost legacy*. Available from https://doi.org/10.1177/1098214014537398

Eisner, E. W. (1975). *The perceptive eye: Toward the reformation of educational evaluation*. Washington: American Educational Research Association.

Eisner, E. W. (1985). *The art of educational evaluation: a personal view*. London: Falmer Press.

Fitzpatrick, J., Sanders, J., & Worthen, B. (2004). *Program evaluation: Alternative approaches and practical guidelines* (3rd ed.). Boston: Allyn & Bacon.

Fitzpatrick, J., Sanders, J., & Worthen, B. (2011). *Program evaluation: Alternative approaches and practical guidelines* (4th ed.). Upper Saddle River: Pearson Education.

Fred, C. L. (2011). Curriculum development: Inductive models. *Schooling, 2*(1), 1–8.

Glaser, B. G., & Strauss, A. L. (1967). *The discovery of grounded theory: Strategies for qualitative research*. Chicago: Aldine.

Heller, A. (2013). *Systemeigenschaft Robustheit - Ein Ansatz zur Bewertung und Maximierung von Robustheit eingebetteter Systeme*. PhD Thesis, Win. Schriftenreihe 'Eingebettete, Selbstorganisierende Systeme' (Vol. 12). Universitätsverlag Chemnitz.

Kaufman, R., Keller, J., & Watkins, R. (2007). What works and what doesn't: Evaluation beyond Kirkpatrick. *Performance and Instruction, 35*(2), 8–12. Available from http://home.gwu.edu/~rwatkins/articles/whatwork.PDF

Khan, B. H. (2004). The People, process and product continuum in e-learning: The e-learning P3 model. *Educational Technology, 44*(5), 33–40. Available from http://asianvu.com/bookstoread/etp/elearning-p3model.pdf

Kirkpatrick, D. L., & Kirkpatrick, J. D. (2006). *Evaluating training programs. The four levels* (3rd ed.). San Francisco: Berrett-Koehler Publishers.

Madaus, G. G., Scriven, M. S., & Stufflebeam, D. L. (1984). Educational evaluation and accountability: A review of quality assurance efforts. *American Behavioral Scientist, 27*(5), 649–673.

Phillips, J. J. (1991). *Handbook of evaluation and measurement methods*. London: Gulf Press.

Provus, M. M. (1969). *The discrepancy evaluation model: An approach to local program improvement and development* (124 pp.). Pittsburgh: Pittsburgh Public Schools.

Rippey, R. M. (1973). *Studies in transactional evaluation*. Berkeley: McCutchan.

Rossi, P. H., & Wright, S. R. (1979). *Evaluation: A systematic approach*. Beverly Hills, CA: Sage.

Ruhe, V., & Zumbo, B. D. (2008). *Evaluation in distance education and E-learning: The unfolding model* (206 pp.). New York: Guilford. ISBN: 978-1-59385-872-8.

Scriven, M. S. (1967). The methodology of evaluation. In R. W. Tyler, R. M. Gagné, & M. Scriven (Eds.), *Perspectives of curriculum evaluation, AERA monograph series on curriculum evaluation* (Vol. 1, pp. 39–81). Chicago: Rand McNally.

Scriven, M. (1973). Goal-free evaluation. In E. R. House (Ed.), *School evaluation: The politics and process* (pp. 319–328). Berkeley: McCutchan.

Scriven, M. S. (1991). Pros and cons about goal-free evaluation. *American Journal of Evaluation, 12*(1), 55–76.

Stufflebeam, D. L. (1968). *Evaluation as enlightenment for decision making*. Columbus: Evaluation Center, Ohio State University.

Stufflebeam, D. L. (1983). The CIPP model for program evaluation. *Evaluation in Education and Human Service, 6*, 117–141.

Stufflebeam, D. L. (1994). Empowerment evaluation, objectivist evaluation, and evaluation standards: Where the future of evaluation should not go and where it needs to go. *American Journal of Evaluation, 15*, 321–338.

Suchman, E. A. (1967). *Evaluative research: Principles and practice in public service and social action programs*. New York: Russell Sage Foundation.

Taba, H. (1962). *Curriculum development: theory and practice*. New York: Harcourt, Brace & World.

Tudevdagva, U., & Hardt, W. (2011). *A new evaluation model for e-learning programs*. Technical Report CSR-11-03, Chemnitz University of Technology.

Tudevdagva, U., & Hardt, W. (2012). A measure theoretical evaluation model for e-learning programs. In *Proceedings of the IADIS on e-Society*, Berlin (pp. 44–52).

Tudevdagva, U., Hardt, W., Tsoy, E. B., & Grif, M. G. (2012). New approach for E-learning evaluation. In *Proceedings of the 7th International Forum on Strategic Technology 2012, Tomsk*, September 17–21, 2012 (pp. 712–715).

Tudevdagva, U., Heller, A., & Hardt, W. (2013a). A model for robustness evaluation of embedded systems. In *Proceedings of the IFOST 2013 Conference, Ulaanbaatar* (pp. 288–292).

Tudevdagva, U., Hardt, W., & Jargalmaa, D. (2013b). The development of logical structures for e-learning evaluation. In *Proceedings of the IADIS on e-learning 2013, Prag, Czech Republic* (pp. 431–435).

Tudevdagva, U., Tomorchodor, L., & Hardt, W. (2014a). The beta version of implementation tool for SURE model. 11th Joint Conference on Knowledge-Based Software Engineering (JCKBSE 2014), Volgograd, September 17–20, 2014. Washington: IEEE Computer Society. In *Journal of Communications in Computer and Information Science*, 466, 243–251.

Tudevdagva, U., Hardt, W., & Bayar-Erdene, L. (2014b). The SURE model for evaluation of complex processes and tool for implementation. In *The 9th International Forum on Strategic Technology (IFOST 2014), Chittagong University of Engineering and Technology, Chittagong*, October 21–23, 2014. Washington, DC: IEEE Computer Society.

Tudevdagva, U., Bayar-Erdene, L., & Hardt, W. (2014c). A self-assessment system for faculty based on the evaluation SURE model. In *Proceedings of The 5th International Conference on Industrial Convergence Technology, ICICT2014*, May 10–11, 2014 (pp. 266–269). Washington, DC: IEEE Computer Society. ISBN 978-99973-46-29-2.

Tudevdagva, U., Jargalmaa, D. & Bayar-Erdene, L., (2014d), Case Study of Structure Oriented Evaluation Model, in *Proceedings of The International Summer School on E-learning, Embedded system and International cooperation, SS2014*, 7–13 July, 2014, Laubusch, Germany (pp. 41–44). ISSN 0947-5125.

Tyler, R. W. (1949). *Basic principles of curriculum and instruction*. Chicago: The University of Chicago Press.

W. K. Kellogg Foundation. (2004). *Logic model development guide: Using logic models to bring together planning, evaluation, and action*. Battle Creek: W. K. Kellogg Foundation.

Warr, P., Bird, M., & Rackham, N. (1970). *Evaluation of management training: A practical framework with cases, for evaluating training needs and results* (112 pp.). London: Gower Press.

Weiss, C. (1998). *Evaluation* (2nd ed.). Englewood Cliffs: Prentice Hall.

Wholey, J. S. (1979). *Evaluation: Promise and performance*. Washington, DC: Urban Institute.

Zhang, W., & Cheng, Y. L. (2012). Quality assurance in E-learning: PDPP evaluation model and its application. *The International Review of Research in Open and Distance Learning, 13*(3), 66–82. Available from http://files.eric.ed.gov/fulltext/EJ1001012.pdf. https://doi.org/10.19173/irrodl.v13i3.1181

Chapter 2
Theory of the Structure Oriented Evaluation Model

2.1 Introduction

An advantage of the structure oriented evaluation model (SURE model) is the data processing part which is adapted to logical structure of process to be evaluated. By taking this structure into account, a multidimensional view to the evaluation process can be achieved (Tudevdagva 2011, 2012). For instance, e-learning is a complex process which consists of several sub processes. Each sub process has its own targets but contribute together in defined manner to achievement of goal of overall process (Tudevdagva et al. 2012, 2013).

Frequently the structural background of a process is ignored at evaluation. As a rule, standard method of statistics is used, which has no relation to the inner structure of process to be evaluated. Mostly linear models with weight factors are applied.

Via weight factors evaluation gets a subjective character, beyond that the linear compromise leads to distortions. If in that context a goal of a sub process was not achieved, which is essential for achieving the overall goal, this can be disguised with a linear approach.

In contrast to this, the SURE model focuses on the logical structure of the process. That is reached by the measure theoretical background of the SURE model. First measure spaces for the single sub processes are considered, which in a second step are connected to a corresponding product space. Thereby a normalized measure is generated, the so-called score by which can be evaluated how a process with given structure achieves its process goal. The obtained score measure complies with the same calculation rules as they hold generally for other normalized measures like probability or reliability measures, for instance (Tudevdagva 2014).

Two logical target structures are at the center of attention: series and parallel structures (Fig. 2.1). For these structures corresponding calculation rules are developed. More complex structures can be evaluated by combination of these rules.

Based on the theoretical scores of a given goal structure estimation values based on checklist data can be obtained, which also refer to the present goal structure. The

U. Tudevdagva, *Structure-Oriented Evaluation*,
https://doi.org/10.1007/978-3-030-44806-6_2

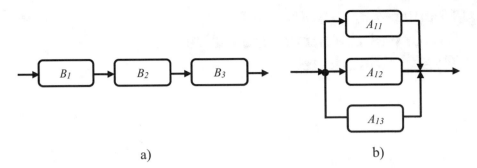

a)   b)

Fig. 2.1 Types of logical structures

estimation formulas depend only on the structure of the evaluation goal and do not need any weight factors. A score value 1 means, process has reached its goal, the value 0 means goal has been failed.

To make scores comparable for different processes or structures a calibration approach is presented. By this method a theoretical evaluation score based on checklist data is transformed in a calibrated empirical evaluation score. The details and properties of this transformation are described in Sect. 2.5. The calibrated empirical evaluation score opens further interpretation possibilities which are helpful for analysis, interpretation, and understanding of evaluation results.

The precision of empirical evaluation scores can be estimated by asymptotic confidence intervals.

2.2 Logical Structures for Evaluation Goals

The e-learning is a complex process which consists of sub processes/elements. Here is defined what to understand formally by an e-learning as "process" in the sense of the SURE model.

Each process consists of several sub process/elements. Elements are partial processes which are involved into the achievement of overall process goal.

Definition Let $\mathcal{G} = \{A_1, \ldots, A_m\}$ be a given set of sub processes where to each element a functional goal is associated. A subset $\mathcal{B} = \{A_1, \ldots, A_r\} \subseteq \mathcal{G}$ of sub processes which interact in sense of the achievement of a defined overall goal is said to be a *process*.

Hence, a random selection of sub processes does not yet form a process, as a rule. The sub processes should work together to achieve overall goal of given complex process. That means, beside the process components a process is characterized by structural properties. These structural properties describe how the involved sub processes contribute to the achievement of overall goal.

The functional structure of a process can be characterized formally by logical operations like AND and OR. In set-theoretical interpretation these are intersection and union. The sub processes of a complex process can interact together in different manner. Depending on how the sub processes of a complex process contribute to the achievement of overall goal two basic structures are distinguished.

Definition A process $B = \{A_1, \ldots, A_r\} \subseteq G$ is a *series process* if the goal of process is achieved if and only if each sub process $A_1, \ldots, A_r$ achieves its goal.

This property of a process can be described formally as follows. Let $B = \{A_1, \ldots, A_r\} \subseteq G$ be a series process consisting of r sub processes $A_1, \ldots, A_r$. Denote by B and $A_1, \ldots, A_r$ the fact that the process B and the sub processes $A_1, \ldots, A_r$ achieve their goals, respectively. Then for a serial process the relation between B and $A_1, \ldots, A_r$ can be formally represented by

$$B = A_1 \cap \cdots \cap A_r = \cap_{i=1}^{r} A_i.$$

That means, a series process B achieves its total goal if all included sub processes achieve their goal. This can be emphasized graphically by a series structure as it is shown in Fig. 1.1a. The order of sub processes in a series structures is of no significance.

Definition A process $B = \{A_1, \ldots, A_s\} \subseteq G$ is a *parallel process* if the goal of process B is achieved if at least one of the sub processes $A_1, \ldots, A_s$ achieves its goal.

Formally then holds

$$B = A_1 \cup A_2 \cup \ldots \cup A_s = \cup_{i=1}^{s} A_i.$$

This can be emphasized graphically by a parallel structure as shown in Fig. 1.1b. The total goal of a parallel process is already achieved if one of the sub processes has reached its goal.

By combination of series and parallel processes more complex process structures can be obtained.

Two processes B_1 and B_2 can overlap.

Definition Two processes $B_1, B_2 \subseteq G$ are said to be *separate*, if

$$B_1 \cap B_2 = \emptyset$$

holds.

Above are given definitions of series and parallel structures in sense of the SURE model. Now the basic concept of process evaluation is considered depending on logical structures of evaluation goals.

Suppose that a process is given which consists of k, $k \geq 1$, sub processes $A_1, \ldots, A_k$, which cooperate together to achieve a defined process goal. The evaluation goal in this case is to evaluate how good this process achieves its process goal.

For this the general measure theory is used (Bauer 2001). The reason to apply general measure theory to calculation rules of the SURE model is because evaluation score of the model should be compatible to the corresponding rules as they are used in established manner in geometrical context at measurement of length, areas, or volumes or in stochastic at calculation of probabilities, for instance. By means of measure theory it becomes possible to evaluate the performance of an e-learning on the same theoretical base as it is done on the fields mentioned above. Subjective influences on the evaluation process like the use of weight factors, for instance, are excluded then.

In sense of measure theoretical description, the evaluation of sub processes $A_1, \ldots, A_k$ has been described in measure theoretical sense by adapted measure spaces. Then, by connection of these spaces to a common product space an evaluation of a process with respect to a given process goal structure becomes possible.

I. Measure Spaces for the Sub Processes For the evaluation how a sub process $A_i \in \{A_1, \ldots, A_k\}$ has achieved its goal a measure space $(\Omega_i, \mathcal{A}_i, Q_i)$ is considered. A measure space consists in sense of general measure theory of three objects Ω_i, $\mathcal{A}_i$, and Q_i which can be defined here as follows:

1. Let $\Omega_i = \{\omega_{i1}, \omega_{i2}\}$ be a two-element set. We denote the elements of this set as *elementary goals*. In this sense the element ω_{i1} is standing for: the process goal of process component A_i has been achieved (is achieved), the element ω_{i2} is standing for: the goal of sub process A_i has not been achieved.
   The set Ω_i is denoted as *goal space* of sub process A_i.
2. Let $\mathcal{A}_i$ be the set of all subsets of goal space Ω_i. Then that is

$$\mathcal{A}_i = \{\emptyset, A_{i1}, A_{i2}, \Omega_i\}.$$

The subsets A_{i1} and A_{i2} are defined by $A_{i1} = \{\omega_{i1}\}$ and $A_{i2} = \{\omega_{i2}\} = \overline{A}_{i1}$, for instance. The elements of the set system $\mathcal{A}_i$ can be interpreted as *goal structures* as follows:

A_{i1} -   goal of sub process A_i has been achieved,
$\overline{A}_{i1}$ -   goal of sub process A_i has not been achieved,
Ω_i -   any goal of sub process A_i has been achieved,
$\emptyset$ -   nothing has been achieved.

The set $\mathcal{A}_i$ is in sense of measure theory a σ-algebra. A σ-algebra is an algebraic structure which is closed against set-theoretical operations with its elements like union, intersection, and complement. That means, set-theoretical or logical operations with the elements of a σ-algebra will not cause any logical confusions.

Here denote the elements of $\mathcal{A}_i$ as *goal structures* and the set $\mathcal{A}_i$ itself as *goal algebra* of process component A_i. In this context: the goal in sense of a goal structure $A \in \mathcal{A}_i$ *has been achieved* (*is achieved*) if an $\omega \in A$ has been observed.

The structure of a goal algebra $\mathcal{A}_i$ is simple and has more of a formal meaning here. The goal algebras are needed in context with a corresponding product space for a common description of process components $A_1, \ldots, A_k$.

3. Let $Q_i : \mathcal{A}_i \rightarrow [0, 1]$ be an additive function from the goal algebra $\mathcal{A}_i$ into the interval $[0, 1]$ where to any given real number $q_i, 0 \leq_i \leq 1$,

$$Q_i(A_{i1}) = q_i, \quad Q_i(\overline{A}_{i1}) = 1 - q_i \quad \text{and} \quad Q_i(\Omega_i) = 1$$

holds. The values of numbers q_i and $1 - q_i$ are or can be interpreted as evaluation scores for the goal structures A_{i1} and $\overline{A}_{i1}$. In this sense the values q_i and $1 - q_i$ form an evaluation distribution over the goal algebra $\mathcal{A}_i$. The function Q_i is a normalized measure on $(\Omega_i, \mathcal{A}_i)$. We denote $Q_i(A_{i1})$ as *score* that the goal of sub process A_i has been achieved, analogously $Q_i(\overline{A}_{i1})$ is the corresponding score that the goal in sense of sub process A_i has not been achieved.

II. The Product Space The triples $(\Omega_i, \mathcal{A}_i, Q_i)$, $i = 1, \ldots, k$, are elementary measure spaces. The obtained spaces can be combined for the sub processes A_i to a product space. By this product space an evaluation of more complex goal structures becomes possible, which include the goals of sub processes $A_1, \ldots, A_k$. For the measure theoretical background used here see Bauer (2001), for instance.

The product space consists again of three objects Ω, $\mathcal{A}$, and Q which are defined now as follows.

1. Let $\Omega = \Omega_1 \times \ldots \times \Omega_k$ be the cross product over the goal spaces $\Omega_1, \ldots, \Omega_k$. The elements of Ω are $\omega = \{\omega_1, \ldots, \omega_k\}$ with $\omega_i \in \Omega_i$ for $i = 1, \ldots, k$. These elements are denoted as *k-dimensional elementary goals* and Ω is then the *k-dimensional goal space*.

2. Let $\mathcal{A}$ be the set of all subsets of k-dimensional goal space Ω. This set of subsets forms a σ-algebra over the goal space Ω. The elements of σ-algebra $\mathcal{A}$ are denoted as *k-dimensional goal structures*. The σ-algebra $\mathcal{A}$ is then the *k-dimensional goal algebra*.

   Some examples of goal structures:

   $A = \{\{\omega_{11}, \omega_{21}, \ldots, \omega_{k1}\}\}$ - all single goals $A_{11}, \ldots, A_{k1}$ in sense of processes $A_1, \ldots, A_k$ have been achieved,

   $B = \{\{\omega_{11}, \ldots, \omega_{k-11}, \omega_{k1}\}, \{\omega_{11}, \ldots, \omega_{k-11}, \omega_{k2}\}\}$ - the single goals $A_{11}, \ldots, A_{k-11}$ have been achieved,

   $C = \{\{\omega_{11}, \omega_{21}, \ldots, \omega_{k-11}, \omega_{k2}\}\}$ - the goals $A_{11}, \ldots, A_{k-11}$ have been achieved, but not the goal A_{k1},

   $D = \{\{\omega_{11}, \omega_{22}, \ldots, \omega_{k2}\}\}$ - only the single goal A_{11} has been achieved,

   $E = \{\{\omega_{11}, \omega_{2i_2}, \ldots, \omega_{ki_k}\}, i_2, \ldots, i_k \in \{1, 2\}\}$ - the single goal $A_{11} \in \mathcal{A}_i$ has been achieved. It holds $E = A_{11} \times \Omega_2 \times \ldots \times \Omega_k$, $A_{11} = \{\omega_{11}\} \in \mathcal{A}_1$.

The last goal structure E is a special goal structure which is directed to the goal of process A_1. The goal algebra $\mathcal{A}$ contains k of such goal structures which are given by

$$A_1 = A_{11} \times \Omega_2 \times \ldots \times \Omega_k, \quad A_{11} = \{\omega_{11}\} \in \mathcal{A}_1, \qquad (2.1)$$

$$A_2 = \Omega_1 \times A_{21} \times \ldots \times \Omega_k, \quad A_{21} = \{\omega_{21}\} \in \mathcal{A}_2,$$

$$\ldots$$

$$A_k = \Omega_1 \times \Omega_2 \times \ldots \times A_{k1}, \quad A_{k1} = \{\omega_{k1}\} \in \mathcal{A}_k.$$

These goal structures are denoted as *single goal structures*.

Besides the single goal structures two further classes of goal structures are of special interest. A goal structure B is said to be a *parallel goal structure* if a subset of single goal structures $A_1, \ldots, A_s \in \mathcal{A}$ exists such that

$$B = \bigcup_{j=1}^{s} A_s$$

holds. A parallel goal structure B achieves its goal if at least one of the included single goal structures $A_1, \ldots, A_s$ reaches its goal. The goal algebra $\mathcal{A}$ contains $2^k - 1$ different parallel goal structures. These are

$$A_1, \ldots, A_k, \quad A_1 \cup A_2, \ldots, A_{k-1} \cup A_k, \quad A_1 \cup A_2 \cup A_3, \ldots$$

$$\ldots, \quad \cup_{i=1}^{k} A_i = \Omega.$$

Because of the set system $\mathcal{A}$ as the set of all subsets of Ω is a σ-algebra it holds:

$$\text{(i)} \quad A_1, \ldots, A_r \in \mathcal{A} \Rightarrow \bigcup_{i=1}^{r} A_i \in \mathcal{A},$$

$$\text{(ii)} \quad A \in \mathcal{A} \Rightarrow \overline{A} \in \mathcal{A}.$$

This implies moreover

$$A_1, \ldots, A_r \in \mathcal{A} \Rightarrow \bigcap_{i=1}^{r} A_i \in \mathcal{A}.$$

So beside union of goal structures and the complement of a goal structure also the intersection of goal structures is again goal structures. That means, the goal algebra $\mathcal{A}$ is logical closed or consistent with respect to application of set-theoretical operations to goal structures.

The parallel goal structures of the following goal structures are of special case. Let $B \in \mathcal{A}$ be a goal structure which to given single goal structures $A_1, \ldots, A_r \in \mathcal{A}$ is defined by

$$B = \bigcap_{i=1}^{r} A_i.$$

Then the goal structure B achieves its goal if and only if when all single goal structures $A_1, \ldots, A_r$ achieve their goal. In analogy to definition of a parallel goal structure, the goal structure is denoted B as a *series goal structure*.

Finally, depending on to which process components two goal structure $B_1, B_2 \in \mathcal{A}$ refer, the goal structures B_1 and B_r are denoted as *separate goal structures* if the associated sets of process components are disjoint.

3. The product measure. Let $Q : \mathcal{A} \rightarrow [0, 1]$ be a map from the goal algebra $\mathcal{A}$ into the interval $[0, 1]$ with the following property. For any goal structure $A = A_1 \times \cdots \times A_k \in \mathcal{A}$ with $A_i \in \mathcal{A}_i, i = 1, \ldots, k$, it holds

$$Q(A) = \prod_{i=1}^{k} Q_i(A_i). \tag{2.2}$$

Then, in sense of measure theory, Q is the product measure of measures Q_i, $i = 1, \ldots, k$. This is according to the Hahn–Kolmogorov theorem of general measure theory a unique defined measure on measurable space $(\Omega, \mathcal{A})$.

Hence by the product measure Q a measure value $Q(A)$ is defined for each goal structure $A \in \mathcal{A}$. The value $Q(A)$ can be interpreted then as an evaluation number for that how the goal in sense of goal structure A has been achieved. In this sense big values $Q(A) \approx 1$ are a hint that the goal in sense of goal structure A has been achieved essentially, whereas a value $Q(A) \approx 0$ is a signal that goal in sense of goal structure A has been failed essentially and $Q(A)$ is denoted as *score for it that the goal of goal structure A has been achieved*.

Collecting this together, the triple $(\Omega, \mathcal{A}, Q)$ forms a corresponding product space which allows an evaluation of all goal structures of goal algebra $\mathcal{A}$ by means of the score measure Q defined by (2.2).

2.3   Calculation Rules for Scores

This subsection describes computation of scores of goal structures. The score Q defined by (2.2) is a normalized measure on $(\Omega, \mathcal{A})$. Each normalized measure possesses the following basic properties.

1. Additivity   According to the addition axiom of measure theory the following rule holds. Let $A_1, \ldots, A_n \in \mathcal{A}$ be pairwise disjoint goal structures such that $A_i \cap A_j = \phi$ for $i \neq j$ and $i, j = 1, \ldots, n$ holds. Then we have

$$Q\left(\bigcup_{i=1}^{n} A_i\right) = \sum_{i=1}^{n} Q(A_i).$$

2. Normalization Rule It holds

$$Q(\Omega) = 1.$$

From these rules further calculation rules are followed. The most important rules are the following. they are standard properties of each normalized measure. They hold correspondingly for the score measure Q considered here.

3. Complement Rule Let $A \in \mathcal{A}$ be an arbitrary goal structure. Then it holds

$$Q(A) = 1 - Q(\overline{A}).$$

Proof This is a standard property of any normalized measure.   □

4. General Addition Rule Let $A_1, \ldots, A_r \in \mathcal{A}$ be arbitrary goal structures. Then it holds

$$Q(A_1 \cup A_2) = Q(A_1) + Q(A_2) - Q(A_1 \cap A_2)$$

as well as

$$Q\left(\bigcup_{i=1}^{r} A_i\right) = \sum_{i=1}^{r} Q(A_i) - \sum_{\substack{i,j=1 \\ i<j}}^{r} Q(A_i \cap A_j) + \sum_{\substack{i,j,k=1 \\ i<j<k}}^{r} Q(A_i \cap A_j \cap A_k)$$

$$- \ldots + (-1)^{r+1} Q(A_1 \cap \cdots \cap A_r).$$

Proof This rule corresponds the general addition rule of measure theory.   □

There are some calculation rules which are of particular interest for evaluation of special goal structures.

5. Product Rule for Series Goal Structures (Series Rule) Let $A_1, \ldots, A_r \in \mathcal{A}$ be single goal structures in sense of relation (2.1) with $Q(A_i) = q_i$, $0 \leq_i \leq 1$, $i = 1, \ldots, r$, $r \leq k$. Let $B = \cap_{i=1}^{r} A_i$ be a corresponding series goal structure. Then it holds

$$Q(B) = Q\left(\bigcap_{i=1}^{r} A_i\right) = \prod_{i=1}^{r} Q(A_i) = \prod_{i=1}^{r} q_i.$$

Proof Without of any loss of generality it can be assumed that the single goal structures $A_1, \ldots, A_r \in \mathcal{A}$ are directed to the first r processes components $A_1, \ldots, A_r$. Then

$$\bigcap_{i=1}^{r} A_i = A_{11} \times \cdots \times A_{r1} \times \Omega_{r+1} \times \cdots \times \Omega_k \quad \text{with} \quad A_{i1} \in \mathcal{A}_i, \ i = 1, \ldots, r.$$

By (2.2) this implies

$$Q\left(\bigcap_{i=1}^{r} A_i\right) = Q_1(A_{11}) \cdots Q_r(A_{r1}) Q_{r+1}(\Omega_{r+1}) \cdots Q_k(\Omega_k). \tag{2.3}$$

For single goal structures A_i holds $Q(A_i) = Q_i(A_{i1})$. With $Q(A_i) = q_i$ for $i = 1, \ldots, r$ and $Q_i(\Omega_i) = 1$ for $i = r + 1, \ldots, k$ this implies

$$Q\left(\bigcap_{i=1}^{r} A_i\right) = \prod_{i=1}^{r} Q_i(A_{11}) = \prod_{i=1}^{r} Q(A_i) = \prod_{i=1}^{r} q_i.$$

$\square$

6. Addition Rule for Parallel Goal Structures (Parallel Rule) Let $A_1, \ldots, A_r \in \mathcal{A}$ be single goal structures with $Q(A_i) = q_i, \ i = 1, \ldots, r, r \leq k$. Let $B = \bigcup_{i=1}^{r} A_i$ be a corresponding parallel goal structure. Then it holds

$$Q(B) = Q\left(\bigcup_{i=1}^{r} A_i\right) - 1 - \prod_{i=1}^{r}(1 - q_i). \tag{2.4}$$

Proof By complement rule and de Morgan's rule holds

$$Q\left(\bigcup_{i=1}^{r} A_i\right) = Q\left(\overline{\overline{\bigcup_{i=1}^{r} A_i}}\right) = 1 - Q\left(\overline{\bigcup_{i=1}^{r} A_i}\right) = 1 - Q\left(\bigcap_{i=1}^{r} \overline{A_i}\right).$$

By product measure property of Q and again by the complement rule holds

$$Q\left(\bigcap_{i=1}^{r} \overline{A_i}\right) = \prod_{i=1}^{r} Q(\overline{A_i}) = \prod_{i=1}^{r}(1 - Q(A_i)) = \prod_{i=1}^{r}(1 - q_i).$$

Collecting this together relation (2.4) is obtained.

A special case is the case $r = 2$. Then it holds

$$Q(A_1 \cup A_2) = 1 - (1 - q_1)(1 - q_2) = q_1 + q_2 - q_1 q_2.$$

For $r = 3$ rule (2.4) modifies to

$$Q(A_1 \cup A_2 \cup A_3) = q_1 + q_2 + q_3 - q_1 q_2 - q_1 q_3 - q_2 q_3 + q_1 q_2 q_3.$$

The addition rule for parallel goal structures is a special case of general addition rule. This is the reason why this rule is denoted here as addition rule too.

7. Product Rule for Generalized Series Goal Structures (Generalized Series Rule) Let $B_1, \ldots, B_r \in \mathcal{A}$ be r, $1 \leq r \leq k$, separate parallel goal structures with

$$B_i = \bigcup_{j=1}^{s_i} A_{ij}, \tag{2.5}$$

generated by s_i single goal structures $A_{ij} \in \mathcal{A}$ with $Q(A_{ij}) = q_{ij}$ for $i = 1, \ldots, r$, $j = 1, \ldots, s_i$, $\sum_{i=1}^{r} s_i = k$. Let $C \in \mathcal{A}$ be a goal structure defined by

$$C = \bigcap_{i=1}^{r} B_i = \bigcap_{i=1}^{r} \bigcup_{j=1}^{s_i} A_{ij}. \tag{2.6}$$

Then it holds

$$Q(C) = Q\left(\bigcap_{i=1}^{r} B_i\right) = Q\left(\bigcap_{i=1}^{r} \bigcup_{j=1}^{s_i} A_{ij}\right) = \prod_{i=1}^{r} Q\left(\bigcup_{j=1}^{s_i} A_{ij}\right)$$

$$= \prod_{i=1}^{r}\left(1 - \prod_{j=1}^{s_i}(1 - q_{ij})\right). \tag{2.7}$$

Proof For each parallel goal structure $B_i = \bigcup_{j=1}^{s_i} A_{ij}$, $i = 1, \ldots, r$, the corresponding product measure space $(\Omega^{(i)}, \mathcal{A}^{(i)}, Q^{(i)})$ generated by the measure spaces $(\Omega_{ij}, \mathcal{A}_{ij}, Q_{ij})$, $j = 1, \ldots, s_i$, is considered. As above, the product measures $Q^{(i)}$ are defined by

$$Q^{(i)}(A_{i1} \times \cdots \times A_{is_i}) = Q_{i1}(A_{i1}) \cdots Q_{is_i}(A_{is_i}), \quad A_{ij} \in \mathcal{A}_{ij}, \ j = 1, \ldots, s_i.$$

Let $A_j^{(i)}$ be a single goal structures of goal algebra $\mathcal{A}^{(i)}$ which is directed to the goal of process component A_{ij} with $Q^{(i)}(A_j^{(i)}) = q_{ij}$, $j = 1, \ldots, s_i$. Then, according to the addition rule for a parallel goal structure $B_i = \bigcup_{j=1}^{s_i} A_j^{(i)} \in \mathcal{A}^{(i)}$ we get

$$Q^{(i)}(B_i) = Q^{(i)}\left(\bigcup_{j=1}^{s_i} A_j^{(i)}\right) = 1 - \prod_{j=1}^{s_i}(1 - q_{ij}). \tag{2.8}$$

If we consider the product of measure spaces $(\Omega^{(1)}, \mathcal{A}^{(1)}, Q^{(1)}), \ldots, (\Omega^{(r)}, \mathcal{A}^{(r)}, Q^{(r)})$, then again the product measure space $(\Omega, \mathcal{A}, Q)$ is obtained which is generated by the measure spaces $(\Omega_{ij}, \mathcal{A}_{ij}, Q_{ij})$ with $i = 1, \ldots, r$ and $j = 1, \ldots, s_i$.

The corresponding product measure Q is then defined as follows. For any goal structure $C = B_1 \times \cdots \times B_r \in \mathcal{A}$, $B_i \in \mathcal{A}^{(i)}$, $i = 1, \ldots, r$, we have

$$Q(C) = Q(B_1 \times \cdots \times B_r) = Q^{(1)}(B_1) \cdots Q^{(r)}(B_r).$$

This, together with (2.8) completes the proof. $\qquad\square$

8. Product Rule for Complex Series Processes (Consistency Rule) Let A'_{ij} be for $i = 1, \ldots, r$ and $j = 1, \ldots, s_i$ separate processes to a given set of process components $\mathcal{G}$. Let C' be a series process whose elements are separate parallel processes B'_i which consist of separate processes $\mathsf{A}'_{i1}, \ldots, \mathsf{A}'_{is_i}$. Denote by C', $B'_1, \ldots, B'_r$ and $A'_{11}, \ldots, A'_{rs_r}$ the corresponding goal structures. Then it holds

$$C' = \bigcap_{i=1}^{r} B'_i = \bigcap_{i=1}^{r} \bigcup_{j=1}^{s_i} A'_{ij}$$

and the score that process C' achieves its goal is

$$Q(C') = \prod_{i=1}^{r} Q(B'_i) = \prod_{i=1}^{r} \left(1 - \prod_{j=1}^{s_i} \left(1 - Q(A'_{ij}) \right) \right). \tag{2.9}$$

Proof This is in case of separateness of included processes A'_{ij} again a property of the product measure Q. $\qquad\square$

This rule is an extension of generalized series rule which allows to include processes instead of sub processes at evaluation of process goals. By this a consistent evaluation of complex process goal structures becomes possible. Hence rule (2.9) can be considered as main rule for process evaluation. By this rule the score calculation of a goal structure C can be reduced consistent to scores for goal structures of separate processes.

The side condition of separateness of involved processes is in many cases fulfilled by the application background. If there are intersections between the involved processes, then it is possible to switch over to separate processes by consideration of finer process structures. This increases the calculation effort a bit. But it is a finite procedure because we have a finite number of process components.

2.4 Estimation of Scores for Goal Structures

Section 2.3 described calculation rules for complex goal structures. This subsection describes calculation rules for computation of estimations scores based on observation data.

Suppose that the considered elementary processes $A_1, \ldots, A_k$ can be observed via ordinal or metrical ordered observation variables $X_1, \ldots, X_k$. Let $\mathcal{X}_i = [x_i', x_i'']$ be the domain or scale of i-th observation variable X_i, $i = 1, \ldots, k$. High values of X_i in the neighborhood of upper bound x_i'' of evaluation scale are an indication of that the goal of process A_i has been achieved, essentially. Small values in the neighborhood of x_i' a corresponding signal that the process goal has been failed, essentially. Observation values $X_i = x_i''$ or $X_i = x_i'$ indicate that the goal of process A_i has been completely achieved or failed, respectively. The scales which are used for observation variables X_i can be continuous or discrete, must be ordered.

Let be

$$C = \bigcap_{i=1}^{r} \bigcup_{j=1}^{s_i} A_{ij}.$$

a given generalized series goal structure. Then the score $Q(C)$ of this goal structure is calculated according to the generalized series rule (2.7) by

$$Q(C) = \prod_{i=1}^{r} \left(1 - \prod_{j=1}^{s_i} (1 - q_{ij}) \right).$$

Let $A_{11}, \ldots, A_{rs_r}$ be the sub processes which contribute to the achievement of goal structure C, let $\vec{X} = (X_{11}, \ldots, X_{rs_r})$ be the vector of observation variables by which the processes $A_{11}, \ldots, A_{rs_r}$ can be observed and let $[x_{ij}', x_{ij}'']$, $x_{ij}' < x_{ij}''$ be the observation interval of variable X_{ij}.

Assume that a sample of size n, $n \geq 1$, to observation vector $\vec{X}$ is given. Such a sample can be obtained, e.g., by a corresponding interrogation of participants of an e-learning program via an assessment checklist after the course is finished. This would be an a-posteriori-interrogation. Or, one could interrogate experts which evaluate the course based on the course materials before the course is held. This would be an a-priori-interrogation.

Let $\vec{x}^{(k)} = (x_{11}^{(k)}, \ldots, x_{rs_r}^{(k)})$ be the kth sampling element in a sample of size n, $k = 1, \ldots, n$. Then at first the sampling values $x_{ij}^{(k)}$ have to be normalized by transforming of these values to the interval $[0, 1]$, the domain for scores $q_{11}, \ldots, q_{rs_r}$. This is reached by the following transformation. Let $q_{ij}^{*(k)}$ for $k = 1, \ldots, n$, $i = 1, \ldots, r$ and $j = 1, \ldots, s_i$ be defined by

$$q_{ij}^{*(k)} = \frac{x_{ij}^{(k)} - x_{ij}'}{x_{ij}'' - x_{ij}'}. \tag{2.10}$$

Then $q_{ij}^{*(k)}$ is an estimation score q_{ij} based on observation value $x_{ij}^{(k)}$. It holds $0 \leq q_{ij}^{*(k)} \leq 1$.

With the so-obtained estimation value for scores $q_{ij}^{*(k)}$ for q_{ij} an estimation value for $Q(C)$ is obtained if formula (2.7) of Sect. 2.3 the unknown scores q_{ij} are substituted by the estimation values $q_{ij}^{*(k)}$. That gives

$$Q^{*(k)}(C) = \prod_{i=1}^{r} \left(1 - \prod_{j=1}^{s_i} \left(1 - q_{ij}^{*(k)} \right) \right) \tag{2.11}$$

for $k = 1, \ldots, n$.

Hence, collecting these values together a sample $(Q^{*(1)}(C), \ldots, Q^{*(n)}(C))$ of size n for score $Q(C)$ is obtained. By the method of moments we can finally obtain via the arithmetic mean an estimation function for the score $Q(C)$ by

$$Q^*(C) = \frac{1}{n} \sum_{k=1}^{n} Q^{*(k)}(C) = \frac{1}{n} \sum_{k=1}^{n} \prod_{i=1}^{r} \left(1 - \prod_{j=1}^{s_j} \left(1 - q_{ij}^{*(k)} \right) \right). \tag{2.12}$$

This is the main formula for estimation of scores for generalized series goal structures based on a sample of size n in context of an interrogation. For a parallel goal structure $B_i = \bigcup_{j=1}^{s_i} A_{ij}$ it can be obtained according to the addition rule for parallel goal structures as estimation function for $Q(B_i)$:

$$Q^*(B_i) = Q^* \left(\bigcup_{j=1}^{s_i} A_{ij} \right) = \frac{1}{n} \sum_{k=1}^{n} \left(1 - \prod_{j=1}^{s_i} \left(1 - q_{ij}^{*(k)} \right) \right). \tag{2.13}$$

For series goal structures $C_i = \bigcap_{j=1}^{r} A_{ij}$ follows from the product rule for series goal structures as estimation score for $Q(C_i)$:

$$Q^*(C_i) = Q^* \left(\bigcap_{i=1}^{r} A_{ij} \right) = \frac{1}{n} \sum_{k=1}^{n} \prod_{j=1}^{r} q_{ij}^{*(k)}. \tag{2.14}$$

The score $q_{ij} = Q(A_{ij})$ of a single goal structure $A_{ij} \in \mathcal{A}$ can be estimated by

$$Q^*(A_{ij}) = \frac{1}{n} \sum_{k=1}^{n} q_{ij}^{*(k)}. \tag{2.15}$$

In case of missing values in the sample the missing values for $q_{ij}^{*(k)}$ can be substituted then by the estimation values $Q^*(A_{ij})$ which are obtained via the arithmetic mean based on the incomplete sample.

2.5 Calibration Rules for Evaluation Scores

Depending on complexity of a logical goal structure the associated score value can be quite small. That can happen even in situations where the scores for involved process components are high. That is a consequence of product measure which forms the base for evaluation of a multidimensional goal structure.

Suppose a series goal structure which consists of five key goals B_i with $Q(B_i) = q = 0.8$ for $i = 1, \ldots, 5$ is given then for the score $Q(C)$ that the aim of goal structure $C = \cap_{i=1}^5 B_i$ is achieved is

$$Q(C) = Q(\cap_{i=1}^5 B_i) = 0.8^5 = 0.3277.$$

Although the key goal scores $Q(B_i)$ are quite high the overall score $Q(C)$ is surprisingly small. If all key goals have an average score $Q(B_i) = 0.5$ for $i = 1, \ldots, 5$, it gives even $Q(C) = 0.5^5 = 0.03125$. That means, the "average level" of a goal structure $C = \cap_{i=1}^5 B_i$ is scored by a comparatively small score value 0.0313. Although this is a general property of any product measure, which follows from the corresponding product rule, such small values can cause irritations especially when the evaluators are not so familiar with measure theoretical background of the SURE model.

Hence a further transformation of estimation score would be desirable by which the following is reached:

1. The obtained score possesses a "normal" value which allows a further interpretation of this value.
2. It becomes possible to judge by the score value whether the aims of a goal structure have been reached in a quality above or below the average.
3. It should be possible to compare score values which belong to different goal structures.

That can be reached by the following transformation or calibration of score value.

Definition Let $C \in \mathcal{A}$ be a goal structure defined by simple goals $A_{ij} \in \mathcal{A}$, $i = 1, \ldots, r$, $j = 1, \ldots, s_i$, such that

$$C = \bigcap_{i=1}^r B_i = \bigcap_{i=1}^r \bigcup_{j=1}^{s_i} A_{ij} \tag{2.16}$$

holds. Let $Q(A_{ij})$ be the score of A_{ij}. Then

$$Q_e(C) = \sqrt[r]{\prod_{i=1}^{r}\left(1 - \sqrt[s_i]{\prod_{j=1}^{s_i}\left(1 - Q(A_{ij})\right)}\right)} \qquad (2.17)$$

is called *evaluation score* (*calibrated score*) of goal structure C.

The evaluation score is a generalized geometric mean. By means the evaluation score a standard score value $Q(C)$ is transformed into a better interpretative score value $Q_e(C)$.

The evaluation score $Q_e(C)$ possesses the following properties.

1. Average Value Rule If $Q(A_{ij}) = 0.5$ holds for $i = 1, \ldots, r$ and $j = 1, \ldots, s_i$, that means all single goals A_{ij} are achieved with average quality, then it holds

$$Q_e(C) = 0.5.$$

This immediately follows from the definitions of $Q(C)$ and $Q_e(C)$.

  If all single goals are achieved with average quality, then a logical goal structure can reach its goal also only in average quality. Hence a value $Q_e(C) = 0.5$ is an indication for evaluator that a goal structure C has achieved its aim in average quality. It is interesting in this context that this average value 0.5 does not depend on the inner structure of a logical goal structure. Hence the value 0.5 becomes a reference value.

  By means of that reference value 0.5 one has a possibility to recognize whether a goal structure has reached its aims better or worse than average. By this it becomes possible to compare different goal structures in this sense too.

  Moreover, if $Q(A_{ij}) = q$ holds for any $0 \leq q \leq 1$, $i = 1, \ldots, r$ and $j = 1, \ldots, s_i$, then it get for a goal structure $C = \bigcap_{i=1}^{r} \bigcup_{j=1}^{s_i} A_{ij}$ according to relation (2.16)

$$Q_e(C) = q.$$

This property can be interpreted as follows. An evaluation score value $Q_e(C) = q$, $0 \leq q \leq 1$, mirrors an "average satisfaction" with the result in sense of goal structure C at the rate of $q \cdot 100\%$. For instance, if an e-learning course, which is evaluated by students with respect to achievement of course goal in sense of a given goal structure with an evaluation value $Q_e(C) = 0.85$ then this value can be interpreted as follows. The average satisfaction level of students with course amounts to 85%.

2. Zero and One Continuity If at least one key goal B_i, $i = 1, \ldots, r$ does not reach its aim, that means it exists at least one index $i^* \in \{1, \ldots, r\}$ such that $Q(B_{i*}) = 0$ holds, then

$$Q_e(C) = Q_e(\bigcap_{i=1}^{r} B_i) = 0.$$

If all scores $Q(B_i) = 1$ for $i = 1, \ldots, r$ then

$$Q_e(C) = Q_e(\cap_{i=1}^r B_i) = 1.$$

3. Monotonicity of Evaluation Score Function Let $C = \cap_{i=1}^r \cup_{j=1}^{s_i} A_{ij}$ be a given goal structure. Let $(q_{11}, \ldots, q_{rs_r})$ and $(q'_{11}, \ldots, q'_{rs_r})$ be two score value sets for $(A_{11}, \ldots, A_{rs_r})$ with

$$q_{ij} \le q'_{ij} \quad \text{for} \quad 1 = 1, \ldots r \quad \text{and} \quad j = 1, \ldots, s_r$$

and let $Q_e(C)$ and $Q'_e(C)$ be the corresponding evaluation scores. Then it holds

$$Q_e(C) \le Q'_e(C).$$

The properties 2 and 3 are direct consequences of the corresponding rules as they hold for the non-calibrated score function Q.

4. The Conservativity of Evaluation Score Function The evaluation score function is conservative in the following sense.

(a) Let $C = B_1 \cap B_1$ be a logical series structure with $Q(B_1) = 0.5 - \varepsilon$ and $Q(B_2) = 0.5 + \varepsilon$ for any $\varepsilon, 0 < \varepsilon < 0.5$. Then we have

$$\frac{1}{2}(Q(B_1) + Q(B_2)) = 0.5.$$

That means, the "average" quality that the goals B_1 and B_2 are achieved is 0.5. The better value in this context is of course the value $Q(B_2) = 0.5 + \varepsilon > 0.5$, not so good is the value $Q(B_1) = 0.5 - \varepsilon < 0.5$, because small values for $Q(B_i)$ can become critical with regard to the evaluation of the entire target structure.

The evaluation score function Q_e behaves itself conservative in such situations. It holds

$$Q_e(C) = \sqrt{Q(B_1)Q(B_2)} = \sqrt{(0.5 - \varepsilon)(0.5 + \varepsilon)} = \sqrt{0.5^2 - \varepsilon^2} < 0.5.$$

Hence, the goal structure $C = B_1 \cap B_2$ is evaluated as below-average which is a right reaction in this case.

(b) Let $C = B_1 \cup B_2$ be a logical parallel structure with $Q(B_1) = 0.5 - \varepsilon$ and $Q(B_2) = 0.5 + \varepsilon$ again. Then in sense of arithmetic mean the "average" quality by which the goals B_1 and B_2 are reached is again 0.5. For a parallel structure the more welcome value is now the value $Q(B_2) = 0.5 + \varepsilon > 0.5$. That means the parallel structure will reach its aim in an "above-average" quality. This is mapped correctly by the evaluation score function. It holds

$$Q_e(C) = Q_e(B_1 \cup B_2) = 1 - \sqrt{(1 - Q(B_1))(1 - Q(B_2))}$$

$$= 1 - \sqrt{(1 - (0.5 - \varepsilon))(1 - (0.5 + \varepsilon))}$$
$$= 1 - \sqrt{0.5^2 - \varepsilon^2} > 0.5.$$

That confirms that the evaluation score function reacts again in the right direction.

From here we can summarize that the evaluation score function Q_e reacts correct as well at logical series as at logical parallel goal structures with respect to deviations of included scores from average value 0.5. This behavior of evaluation score function Q_e we denote as "conservative" behavior of Q_e against deviations of included scores from the average value 0.5. This conservative behavior holds analogously if we consider deviations from a score value $q_0 \neq 0.5$ with $Q(B_1) = q_0 - \varepsilon$ and $Q(B_2) = q_0 + \varepsilon$.

5. Estimation of Evaluation Scores In analogous manner like at estimation of score of a goal structure based on observation results of an adapted checklist the evaluation score can be estimated as follows.

Suppose a sample of size n of corresponding checklist results is given. Let according to formula (2.10) of Sect. 2.4 $q_{ij}^{*(k)}$ be for $i = 1, \ldots, r, j = 1, \ldots, s_i$ and $k = 1, \ldots, n$ the estimation value for $Q(A_{ij})$ based on the kth checklist data record. Then model obtains for each checklist record an estimation value for $Q_e(C)$ with

$$C = \bigcap_{i=1}^{r} \bigcup_{j=1}^{s_i} A_{ij}$$

if we substitute in formula (2.17) the unknown score values $Q(A_{ij})$ by the estimation values $q_{ij}^{*(k)}$. For each of the n data records of our checklist sample model gets an estimation value for $Q_e(C)$ by

$$Q_e^{*(k)}(C) = \sqrt[r]{\prod_{i=1}^{r} \left(1 - \sqrt[s_i]{\prod_{j=1}^{s_i} \left(1 - q_{ij}^{*(k)} \right)} \right)}, \qquad (2.18)$$

$k = 1, \ldots, n$. By the method of moments model gets then via the arithmetic mean over the values $Q_e^{*(1)}(C), \ldots, Q_e^{*(n)}(C)$ an estimation value for evaluation score $Q_e(C)$.

Definition To a given goal structure

$$C = \bigcap_{i=1}^{r} \bigcup_{j=1}^{s_i} A_{ij}$$

and a sample of size n of checklist results the value

$$Q_e^*(C) = \frac{1}{n}\sum_{k=1}^{n} Q_e^{*(k)}(C) = \frac{1}{n}\sum_{k=1}^{n} \sqrt[r]{\prod_{i=1}^{r}\left(1 - \sqrt[s_i]{\prod_{j=1}^{s_i}\left(1 - q_{ij}^{*(k)}\right)}\right)} \qquad (2.19)$$

is denoted as *empirical evaluation score* for goal structure C.

This is the main formula for estimation of evaluation scores of goal structures.

6. Confidence Intervals for Evaluation Scores Checklist results depend always on random influences which work in the background. Hence a checklist result like $(Q_e^{*(1)}(C), \ldots, Q_e^{*(n)}(C))$ can be considered as a sample of size n for the evaluation score $Q_e(C)$. In this sense the evaluation score $Q_e(C)$ is a random variable. Then like any random variable the evaluation score $Q_e(C)$ has an expectation value as well as a variance. Denote by

$$\tilde{q} = E Q_e(C) \quad \text{and} \quad \tilde{\sigma}^2 = D^2 Q_e(C)$$

the expectation value and the variance of $Q_e(C)$, respectively. Depending on the value of variance $\tilde{\sigma}^2$ an estimation value $Q_e^*(C)$ for $Q_e(C)$ is more or less precise. To evaluate the precision of an evaluation score value evaluator can consider a confidence interval for the expectation value $\tilde{q}$ of evaluation score $Q_e(C)$.

A confidence interval for $\tilde{q}$ is a random interval which covers the unknown score value $\tilde{q}$ with a pre-given probability, the confidence level $1 - \alpha$. Frequently used values for the probability α are values like $\alpha = 0.1, 0.05$ or 0.01. The construction of a confidence interval requires the knowledge of probability distribution of observation variable $Q_e(C)$. This distribution is because of the complexity of random background in the present case not known. Therefore we will consider an asymptotic confidence interval. This is a possibility nevertheless to get a confidence interval if the sample size is sufficiently large.

Model proceeds as follows. By the central limit theorem the empirical evaluation score

$$Q_e^*(C) = \frac{1}{n}\sum_{k=1}^{n} Q_e^{*(k)}(C)$$

is asymptotically, that means for $n \to \infty$, normal distributed with mean $\tilde{q}$ and variance $\tilde{\sigma}^2/n$. This implies the standardized random variable

$$Z_n = \sqrt{n}\,\frac{Q_e^*(C) - \tilde{q}}{\tilde{\sigma}}$$

is asymptotically $N(0, 1)$-distributed.

A confidence interval for $\tilde{q}$ can be obtained then as follows. Let z_α^* be defined to a pre-given probability α, $0 < \alpha < 1$, by

$$\Phi(z_\alpha^*) = \int_{-\infty}^{z_\alpha^*} \frac{1}{\sqrt{2\pi}} \exp\left(-\frac{z^2}{2}\right) dz = 1 - \alpha/2.$$

Here $\Phi(z)$ denotes the distribution function and z_α^* the $(1 - \alpha/2)$-quantile of standardized normal distribution. Then it holds with probability $1 - \alpha$ for sufficient large values of sample size n

$$-z_\alpha^* \leq Z_n \leq z_\alpha^*$$

or

$$Q_e^*(C) - z_\alpha^* \frac{\tilde{\sigma}}{\sqrt{n}} \leq \tilde{q} \leq Q_e^*(C) + z_\alpha^* \frac{\tilde{\sigma}}{\sqrt{n}}.$$

Definition The interval

$$\left[Q_e^*(C) - z_\alpha^* \frac{\tilde{\sigma}}{\sqrt{n}}, \ Q_e^*(C) + z_\alpha^* \frac{\tilde{\sigma}}{\sqrt{n}} \right]$$

is said to be an *asymptotic two-sided confidence interval* for the average evaluation score value $\tilde{q}$ at *confidence level* $1 - \alpha$.

This interval is to interpret statistically as follows. It covers in average in $(1 - \alpha) \cdot 100\%$ of all cases the unknown parameter value $\tilde{q}$.

The problem here is that the variance $D^2 Q_e(C) = \tilde{\sigma}^2$ is unknown. The usual approach in such cases is to substitute the unknown variance $\tilde{\sigma}^2$ by a corresponding estimation value for the unknown variance like

$$S^2 = \frac{1}{n-1} \sum_{k=1}^{n} \left(Q_e^{*(k)}(C) - Q_e^*(C) \right)^2 \quad \text{with} \quad Q_e^*(C) = \frac{1}{n} \sum_{k=1}^{n} Q_e^{*(k)}(C).$$

The so-obtained interval

$$\left[Q_e^*(C) - z_\alpha^* \frac{S}{\sqrt{n}}, \ Q_e^*(C) + z_\alpha^* \frac{S}{\sqrt{n}} \right]$$

is then an approximation for a two-sided asymptotic confidence interval for the evaluation score $\tilde{q}$ at confidence level $1 - \alpha$ based on a sample of size n.

Beside a confidence interval the sample standard deviation

$$\tilde{\sigma}^* = \sqrt{\frac{1}{n-1} \sum_{k=1}^{n} \left(Q_e^{*(k)}(C) - Q_e^*(C) \right)^2}$$

is a statistical quantity describing the precision of an score estimation value.

2.6 Summary

This section describes the theoretical background of the SURE model. Basic idea of this model is the use of general measure theory for evaluation of structural processes.

Section starts with a structural analysis of processes consisting of several sub processes which cooperate together in a defined manner to reach a given process goal. Of particular meaning are series and parallel sub processes structures with the help of which very complex process structures can be described.

The measure theoretical description of a structured process starts with definition of elementary measure spaces for evaluation of involved sub processes. These measure spaces are base for a further product space by which a multidimensional view to evaluation is reached. The so-obtained product measure is then a possibility to measure or to score how a given process will reach its goal. The score measure satisfies thereby the same calculation rules as any normalized measure where the series and parallel rule are of particular importance.

On the basis of the obtained score calculation rules a statistical method for estimation of score values is developed using checklist data to get corresponding empirical scores, i.e. observation-based scores.

Depending on complexity of considered process to be evaluated, score values can become quite small. Therefore a calibration method for score values is developed which is helpfully for further interpretation of evaluation results. By means of an adapted statistical method for given checklist data corresponding estimation values can be obtained, the so-called empirical evaluation scores.

The precision of empirical scores can be estimated by means of asymptotic confidence intervals.

References

Bauer, H. (2001). *Measure and integration theory. De Gruyter studies in mathematics*. Berlin: De Gruyter.

Tudevdagva, U. (2014). *Structure oriented evaluation model for E-learning*. Wissenschaftliche Schriftenreihe Eingebettete Selbstorganisierende Systeme, Universitätsverlag Chemnitz, Chemnitz, July 2014 (123 pp.). ISBN: 978-3-944640-20-4, ISSN: 2196-3932.

Tudevdagva, U., & Hardt, W. (2011). *A new evaluation model for e-learning programs*. Technical Report CSR-11-03, Chemnitz University of Technology.

Tudevdagva, U., & Hardt, W. (2012). A measure theoretical evaluation model for e-learning programs. In *Proceedings of the IADIS on e-Society, Berlin* (pp. 44–52).

Tudevdagva, U., Hardt, W., Evgeny, T., & Grif, M. (2012). New approach for E-learning evaluation. In *Proceedings of the 7th International Forum on Strategic Technology, Tomsk* (pp. 712–715).

Tudevdagva, U., Hardt, W., & Jargalmaa, D. (2013). The development of logical structures for e-learning evaluation. In *Proceedings of the IADIS on e-Learning 2013, Prag* (pp. 431–435).

Chapter 3
Evaluation Examples of the SURE Model

3.1 An Example for Illustration of Data Processing

A highlight of the SURE model is its structure oriented data processing part, which is illustrated here using example. The following describes data processing rules of SURE model based on an abstract example. The example focuses only on data processing part in detail. All other steps of eight steps will be just mentioned as necessary steps of model.

Step 1. Definition of key goals

Consider a key goal structure with three key goals (Fig. 3.1).

Series structure of key goals indicates that all key goals have to reach their targets. Only under that condition the final evaluation will be successful (Tudevdagva 2014).

Step 2. Definition of sub goals

Assume that first and third key goals have no further sub goals and second key goal consists of two sub goals (Fig. 3.2).

Step 3. Confirmation of goal structures

Suppose that defined goal structure is accepted by all involved groups of evaluation and has been confirmed by an official protocol.

Step 4. Creation of checklist

This step is not further discussed here, because this example serves the understanding of data processing part.

Step 5. Confirmation of checklist

Suppose that the created checklist is accepted by involved groups of evaluations and has been confirmed by an official protocol.

Step 6. Data collection

In case of a real evaluation a confirmed checklist has to be send to participants of interrogation. Their answers form checklist data.

For data processing example simulated data are used, obtained by online tool of the SURE model (Fig. 3.3). To generate random data the corresponding structure data have to be entered. These are:

© Springer Nature Switzerland AG 2020
U. Tudevdagva, *Structure-Oriented Evaluation*,
https://doi.org/10.1007/978-3-030-44806-6_3

Fig. 3.1 The structure of key
goals

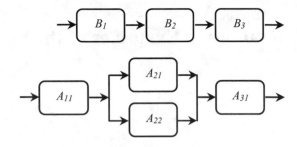

Fig. 3.2 The structure of sub
goals

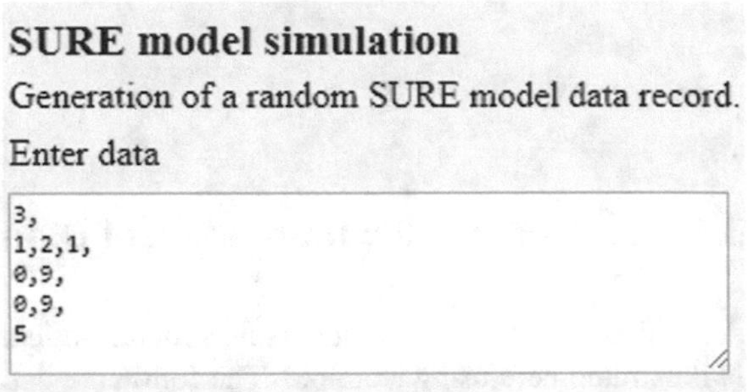

Fig. 3.3 The window for
data input

SURE model simulation

Generation of a random SURE model data record.

Enter data

```
3,
1,2,1,
0,9,
0,9,
5
```

3,   number of key goals,

1,2,1,   numbers of sub goals of considered key goals,

0,9,   boundaries of evaluation interval,

0,9,   boundaries for random evaluation values,

5,   desired sample size.

These values have to be entered into input window of online tool for data simulation (Fig. 3.3). Figure 3.4 shows a corresponding simulation result.

These are the simulation data records.

$$3, 4, 0, 9,$$

$$9, 0, 0, 9,$$

$$0, 7, 5, 8,$$

$$2, 0, 4, 4,$$

$$3, 5, 6, 4,$$

The single values are realizations of an uniformly distributed random variable with values in the set $\{0, 1, \ldots, 9\}$.

Step 7. Data processing

Now data processing can be started according to the rules described in Sect. 3.2.

First step of data processing is always normalization of observation data, see Formula (10) from Sect. 3.2. For example, first participant of interrogation has

Fig. 3.4 Data record with
simulated data

SURE model simulation

Simulation data record

```
3,
1,2,1,
0,9,
3,4,0,9,
9,0,0,9,
0,7,5,8,
2,0,4,4,
3,5,6,4,
```

Fig. 3.5 Table with
normalized data

		B_1	B_2		B_3
k		A_{11}	A_{21}	A_{22}	A_{31}
1		0.33	0.44	0	1
2		1	0	0	1
3		0	0.78	0.56	0.89
4		0.22	0	0.44	0.44
5		0.33	0.56	0.67	0.44

evaluated how goal A_{11} is reached by the value 3. Hence $x_{11}^{(1)} = 3$. With $x_{11}' = 0$
and $x_{11}'' = 9$ this implies

$$q_{11}^1 = \frac{x_{11}^1 - x_{11}'}{x_{11}'' - x_{11}'} = \frac{3-0}{9-0} = 0.33.$$

This is to repeat analogously for all other observation values $x_{ij}^{(k)}$. After that all
scores $q_{ii}^{(k)}$ should have values between 0 and 1. The normalized data are shown in
Fig. 3.5.

After normalization the empirical scores $Q^{*(k)}(C)$ have to be calculated by
Formula (11) for all checklist data records, $k = 1, \ldots, 5$. For example, for first
data record $x_{11}^{(1)}, x_{21}^{(1)}, x_{22}^{(1)}, x_{31}^{(1)}$ holds

$$Q_1^*(C) = \left(1 - \prod_{j=1}^{1}\left(1 - q_{1j}^{*(1)}\right)\right)\left(1 - \prod_{j=1}^{2}\left(1 - q_{2j}^{*(1)}\right)\right).$$

$$\cdot \left(1 - \prod_{j=1}^{1}\left(1 - q_{3j}^{*(1)}\right)\right)$$

$$= (1 - (1 - 0.33)) \cdot (1 - (1 - 0.44)(1 - 0)) \cdot (1 - (1 - 1))$$
$$= 0.15.$$

This is the empirical score for first data record. Analogously the remaining scores are to be calculated as:

$$Q_1^*(C) = 0.15,$$
$$Q_2^*(C) = 0,$$
$$Q_3^*(C) = 0,$$
$$Q_4^*(C) = 0.04,$$
$$Q_5^*(C) = 0.13.$$

Now the empirical score $Q^*(C)$ for the whole sample can be computed as an arithmetic mean over the obtained empirical scores of data records. It holds

$$Q^*(C) = \frac{1}{5} \sum_{k=1}^{5} Q_k^*(C) = 0.0636.$$

Depending on complexity of evaluation goal structure this value can become quite small because of product structure of general score calculation rule, which is also shown in the example. Therefore, additionally to empirical score $Q^*(C)$ the SURE model offers a further calibrated score parameter, the so-called empirical evaluation score, see Sect. 4.2. This parameter is calculated according formula:

$$Q_e^*(C) = \frac{1}{n} \sum_{k=1}^{n} Q_e^{*(k)}(C),$$

where $Q_e^{*(k)}(C)$ is given by (2.18) for $k = 1, \ldots, n,$. For $k = 1$ this implies

$$Q_e^{*(1)}(C) = \sqrt[3]{\prod_{i=1}^{3} \left(1 - \sqrt[s_i]{\prod_{j=1}^{s_i} \left(1 - q_{ij}^{*(1)} \right)} \right)}$$

$$= \sqrt[3]{\left(1 - \sqrt[1]{(1 - 0.33)} \right) \left(1 - \sqrt[2]{(1 - 0.44)(1 - 0)} \right) \left(1 - \sqrt[1]{(1 - 1)} \right)}$$

$$= 0.44$$

as well as

$$Q_e^{*(2)}(C) = 0,$$

$$Q_e^{*(3)}(C) = 0,$$

$$Q_e^{*(4)}(C) = 0.29,$$

$$Q_e^{*(5)}(C) = 0.45.$$

From this follows

$$Q_e^*(C) = \frac{1}{5} \sum_{k=1}^{5} Q_e^{*(k)}(C) = 0.2365.$$

This value can be interpreted as follows. According to the given checklist data the goal of considered process has been achieved with an average satisfaction level of 23.6%. Background of that interpretation is the following: If all questions of checklist are evaluated with $0.2365 \cdot 9 = 2.1285$, then this empirical evaluation score is obtained, see Sect. 3.2.

Similarly, further empirical scores $Q^*(B_i)$ and $Q_e^*(B_i)$, $i = 1, \ldots, 3$ can be computed. For results see Fig. 3.6.

Step 8. Evaluation report

Evaluation report covers all obtained empirical evaluation scores in sense of involved groups.

Evaluation table (empirical evaluation scores and standard scores)

	B_1	B_2		B_3		
k	A_{11}	A_{21}	A_{22}	A_{31}	$Q_{e,k}^*(C)$	$Q_k^*(C)$
1	0.33	0.44	0	1	0.44	0.15
2	1	0	0	1	0	0
3	0	0.78	0.56	0.89	0	0
4	0.22	0	0.44	0.44	0.29	0.04
5	0.33	0.56	0.67	0.44	0.45	0.13
$Q^*(A_{ij})$	0.38	0.36	0.33	0.76	$Q_e^*(C) = 0.2365$	
$Q_e^*(B_i)$	0.38	0.36		0.76		
$Q^*(B_i)$	0.38	0.53		0.76		$Q^*(C) = 0.0636$

Evaluation interval $[x_0, x_1] = [0, 9]$

Fig. 3.6 Result view of online data processing tool in table

3.2   E-learning Example with Simulated Data

The SURE model was originally developed to evaluate the performance of e-learning (Tudevdagva and Hardt 2011, 2012; Tudevdagva et al. 2012, 2013a,b, 2014a,c,b,d). Therefore, this example is about e-learning. E-learning is a very complex process involving many different groups (Englisch et al. 2019). The SURE model is a way to evaluate structure oriented e-learning with a well-founded formal background for data processing.

Assuming an evaluator is given the task to evaluate an e-learning course. Then the first step in this context is always a task analysis where answers to the following questions should be found:

- What is the aim of evaluation?
- Who is interested in the result of evaluation?
- Who is involved into development and implementation of e-learning course?
- To whom is the e-learning course addressed?
- What is the view and expectations of involved groups to evaluation process?
- What do they want to figure out via evaluation?

The aim of evaluation is to assess the performance of the course under review.

The following groups were involved in the development of the course: stakeholders, professors, and multimedia developers. In addition, the administration stuff, professors, and students participated in the implementation. Evaluation results are of interest for all involved groups. As a rule, there is also a public interest and an interest from other educational institutions. The course was aimed at a broad audience, not at students only, it was a free course for adults with an interest in the subject.

The views on evaluation are very different. Stakeholders usually understand this as a kind of monitoring process that reflects the value of their investments and provides information for further decisions. The professors see evaluation as a way of measuring the performance of e-learning and try to identify weaknesses for improving the course. For them, the feedback of the students on the course is also important. The interest of multimedia or technical staff is to find out what kind of course elements are important for learners in e-learning, what kind of media and digital materials contribute to improving the quality of an e-course.

For administrators, evaluation means the evaluation of their service to learners and professors. It is interesting for them to find out how well the administrative elements work in the e-learning course, such as registration, certificate application and communication in case of administrative problems via the learning platform.

Students and learners are interested in giving professors, universities, and stakeholders their own feedback through an evaluation process. The evaluation of e-learning is one of the best ways to convey messages about the development and implementation of e-learning to the other groups involved, and it is the main understanding and perspective of the learner on the evaluation.

Step 1. Definition of key goals.

Evaluation process by SURE model starts with definition of an evaluation goal. In the example considered, it is assumed that five groups are involved in e-learning and the evaluator must first identify their expectations of the evaluation objectives.

Based on the given five involved groups the evaluator defines five key goals: Accessibility, Content, Goals and objectives, Interactivity, and Visual design (Fig. 3.7).

The number of key goals may be higher depending on the number of expectations of the groups involved. In the SURE model, the key goals are visualized using a logical series structure (Fig. 3.8).

Main characteristic of key goal structure is that key goals are connected to each other in sense of a logical AND. That means, the goal structure will be evaluated positive or successful only, if all key goals achieve their target. For example, if any of the defined five key goals is evaluated as failed (evaluation score is zero, for instance), then overall goal is considered as not achieved and evaluation score gets the value zero, too. Therefore, evaluator has to define very carefully the key goals of evaluation. The sequence and numbering of the key targets in the logical structure are irrelevant in this context and do not play any role in further score calculation. The result of first step should be the defined key goals and the so-called fixed overall goal of evaluation.

Step 2. Definition of sub goals.

Based on the defined key goals further sub goals of these key goals can be considered if that is necessary or desired or is given by the actual overall goal structure. Frequently, key goals have no sub goals and a goal structure can consist of key goals only. Thereby it should be noted that the goals of the agreed key goals can be achieved independently from each other. In the example considered here it is assumed that the evaluator has identified some sub goals at five key goals. Aim of sub goals is to contribute to achievement of a key goal. If a key goal has no sub goals, then it can be considered formally as a special single sub goal, too.

Figure 3.9 shows definition of sub goals. The first key goal "Accessibility" is identical to a single sub goal: "Navigation of course." The second key goal consists of two sub goals. "Relevant level of content to learners' expectation" and "Completeness of content" together belong to achievement of this key goal. If

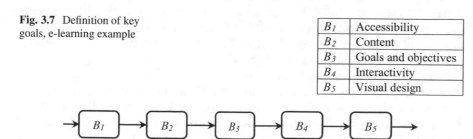

Fig. 3.7 Definition of key goals, e-learning example

B_1	Accessibility
B_2	Content
B_3	Goals and objectives
B_4	Interactivity
B_5	Visual design

Fig. 3.8 The logical structure of defined key goals, e-learning example

Key goals	Sub goals	Definition
B_1	A_1	Navigation of course
B_2	A_{21}	Relevant level of content to learners' expectation
	A_{22}	Completeness of content
B_3	A_{31}	Clearness of course objective
	A_{32}	Fitness of content to course objectives
	A_{33}	Learners performance
B_4	A_{41}	Interactivity possibilities
B_5	A_{51}	Visual appealing
	A_{52}	Effective color design
	A_{53}	Effective font size
	A_{54}	Overall visual design

Fig. 3.9 Definition of sub goals, e-learning example

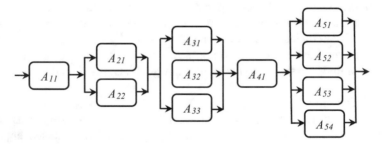

Fig. 3.10 The logical structure of defined sub goals, e-learning example

targets of these two sub goals are achieved or one of these goals is achieved, then by model the target of second key goal is achieved. The third key goal includes three sub goals. "Clearness of course objective," "Fitness of content to course objectives," and "Learners performance" contribute to achievement of this key goal. The fourth key goal consists of again a single sub goal: "Interactivity possibilities." If this single goal fails its aim, fourth key goal fails the target. The fifth key goal consists of four sub goals: "Visual appealing," "Effective color design," "Effective font size," and "Overall visual design" contribute to achievement of this key goal.

The key goal structures B_1 and B_4 are single goal structures, the key goal structures B_2, B_3, and B_5 are parallel structures. The logical structure of this structure of evaluation goal is shown by Fig. 3.10. Result of second step should be definition and confirmation of sub goals.

Step 3. Confirmation of goal structures

In this step the evaluator has to discuss with all involved groups the defined goals. Involved groups have to analyze the defined goals and have to give their feedback to evaluator. If groups have other views to the evaluation goals, evaluator should repeat first two steps and analyze again the defined goals. After updating the goal definitions, it should be discussed again with all involved groups. Evaluator can go over to next step if all involved groups accept the defined goal structure.

Sub goals	Criteria for data collection	0	1	2	3	4	5	6	7	8	9
		Strongly disagree							Strongly agree		
A_1	It was easy to understand how to navigate the course										
A_{21}	Content of course met my expectations										
A_{22}	Course content was complete										
A_{31}	Course objectives were clear for me										
A_{32}	Contents followed the objectives of course										
A_{33}	Objectives of course were tied closely to the performance context of learners										
A_{41}	Easy and interactive features										
A_{51}	Course was appealing										
A_{52}	Color combination was excellent										
A_{53}	Font sizes were selected correctly, easy to read										
A_{54}	Visual design of course was excellent										

Fig. 3.11 Checklist for data collections, e-learning example

The outcome of this third step should be confirmed by an agreement on logical structure of evaluation goals with signatures of representatives of involved groups.

Step 4. Creation of checklist

During this step the evaluator has to prepare the checklist or questionnaire for data collection. The checklist has to be developed based on logical structure of evaluation goals. Defined sub goals form the basis for an adapted checklist for data collection. In Fig. 3.11 checklist is shown for e-learning evaluation considered here.

Step 5. Confirmation of checklist

In this step the evaluator should send created checklist to all involved groups for confirmation. Each involved group has to test checklist and if there are any criteria to improve the formulation then they have to give feedback to the evaluator. Important is, checklist may not deviate from pre-given logical structure, it must be adapted to the logical structure of evaluation goals. But necessary, the corresponding questions can still be modified by the involved groups.

The outcome of this fifth step should again be an official document about agreement on the created checklist for data collection, again with signatures of the representatives of involved groups.

Step 6. Data collection

Data collection for evaluation can be done in different ways. Traditionally, paper-based version is an accepted method for data collection. Moreover, nowadays many commercial and non-commercial web sites offer online tools which can be used for data collection. The evaluator has to decide that in accordance with involved groups.

For examples considered here random generated data are used. Assume that based on an adapted checklist a sample of size $n = 15$ has been obtained. The evaluation intervals are for all questions equal. The evaluation scale or set includes the numbers $0, 1, \ldots, 9$. Zero means the goal has been failed, the value 9 stands for the goal has been fully achieved.

For better explanation of the SURE model three different examples with simulated data are presented:

- The satisfied students
- The unsatisfied students
- The gambling students

Example 3.2.1 The satisfied students.

To generate a random observation data set for this example the online tool for the SURE model was used, see Sect. 3.4.

The simulation of a SURE model checklist data sample requires certain simulation parameters. That is the simulation parameter set for data simulation in case of "satisfied" students:

$$5,$$

$$1, 2, 3, 1, 4,$$

$$0, 9,$$

$$4, 9,$$

$$15.$$

These parameters have the following meaning:

5,	–	number of key goals,
1,2,3,1,4,	–	numbers of associated sub goals,
0,9,	–	bounds of evaluation interval,
4,9,	–	interval for random selection of an answer, answers are selected by means of a random number generator, who generates uniformly distributed random numbers from set {4, 5, ...9}—all answers are ≥ 4, that means, students are quite sufficient,
15	–	sample size, number of evaluation data records.

The simulation parameter above has to be entered into input window of online simulator (http://uranchimeg.com/sure/eva_simul.html).

The output result, the generated simulation data set, of online tool is shown in Fig. 3.12.

Example 3.2.2 The unsatisfied students.

Generating random data under this aspect the interval for answers is changed to [0, 5]. That means, all questions are answered with a score value from set 0, 1, . . . , 5. The simulation parameter set for simulator of SURE model tool is then:

k	B_1 A_{11}	B_2 A_{21}	A_{22}	B_3 A_{31}	A_{32}	A_{33}	B_4 A_{41}	B_5 A_{51}	A_{52}	A_{53}	A_{54}
1	9	9	9	5	6	9	9	5	8	9	5
2	5	9	5	5	7	6	7	4	4	6	6
3	9	7	4	5	7	9	5	4	9	5	4
4	8	7	6	8	6	8	6	6	4	8	5
5	5	4	9	8	7	9	8	9	5	8	7
6	5	4	4	5	6	5	4	7	5	9	4
7	7	7	7	6	4	9	6	8	5	8	9
8	4	7	6	9	6	6	5	4	9	7	5
9	9	8	7	5	9	5	6	9	5	4	7
10	9	6	7	8	9	6	9	8	5	4	5
11	7	4	7	4	5	8	9	8	4	9	7
12	7	4	7	9	6	6	4	7	9	4	9
13	7	8	9	9	8	8	5	9	4	9	9
14	7	9	4	6	9	9	6	8	7	4	9
15	4	9	6	7	4	9	7	4	8	4	9

Fig. 3.12 Simulated data for satisfied students

$$5,$$
$$1, 2, 3, 1, 4,$$
$$0, 9,$$
$$0, 5,$$
$$15.$$

Table in Fig. 3.13 contains the generated checklist data.

Example 3.2.3 The gambling students.

In this case, the response set includes all values of set $\{0, 1, \ldots, 9\}$. This means that a value is randomly selected as the answer from the set of all possible answers and the following simulation parameter set is used for simulation:

$$5,$$
$$1, 2, 3, 1, 4,$$
$$0, 9,$$
$$0, 9,$$
$$15.$$

Figure 3.14 shows a corresponding simulation data set.

	B_1	B_2		B_3			B_4	B_5			
k	A_{11}	A_{21}	A_{22}	A_{31}	A_{32}	A_{33}	A_{41}	A_{51}	A_{52}	A_{53}	A_{54}
1	5	4	3	5	5	4	4	1	0	5	5
2	4	0	5	3	1	5	4	5	4	3	4
3	1	3	2	1	2	3	0	4	5	0	2
4	3	0	2	1	4	3	2	4	3	1	4
5	2	4	0	1	3	5	0	0	3	1	3
6	0	3	0	3	4	5	3	5	1	0	5
7	4	2	4	1	5	2	5	0	0	1	4
8	1	3	1	0	4	2	4	0	5	4	4
9	0	1	2	5	5	2	1	5	1	5	2
10	5	1	1	2	0	1	3	2	0	4	0
11	2	4	5	4	2	5	4	1	3	4	2
12	0	4	1	2	5	1	4	5.	4	4	0
13	0	0	1	2	4	4	3	2	4	5	1
14	3	3	3	3	2	4	0	1	1	0	5
15	3	3	5	5	2	4	4	0	5	5	1

Fig. 3.13 Simulated data for unsatisfied students

	B_1	B_2		B_3			B_4	B_5			
k	A_{11}	A_{21}	A_{22}	A_{31}	A_{32}	A_{33}	A_{41}	A_{51}	A_{52}	A_{53}	A_{54}
1	5	9	7	9	9	4	5	2	2	8	9
2	6	0	0	7	3	3	5	2	6	2	8
3	3	6	9	1	6	2	5	1	2	0	1
4	9	9	0	4	5	3	7	3	2	3	3
5	3	0	7	6	6	9	2	8	8	6	5
6	7	8	1	0	3	3	3	3	4	2	3
7	5	7	8	8	4	2	1	8	6	4	8
8	3	1	5	3	3	3	1	0	8	9	9
9	0	9	2	3	2	6	7	5	0	2	3
10	8	0	8	1	1	6	7	6	5	1	7
11	0	5	0	4	6	1	3	6	0	3	6
12	3	6	8	0	3	4	0	5	8	9	6
13	6	0	7	3	8	3	8	9	1	9	5
14	2	3	1	3	7	8	4	0	4	8	6
15	2	8	0	7	8	6	6	8	3	2	8

Fig. 3.14 Simulated data for gambling students

Step 7. Data processing

Data processing should be performed according to the calculation rules in Sect. 3.2. For this the evaluator can use the SURE model online tool. First to enter are the corresponding structure parameters and then the simulation data as comma separated file (CSV file, Fig. 3.15).

There are various options for displaying the results. To recognize tendencies in the evaluation data records the data can be colored and supported with color scales. The checklist data can be displayed in normalized or non-normalized form. The standard form of the result display contains the empirical evaluation scores. Optionally, the empirical standard scores can also be displayed. After selection of options for result display and pressing the **SEND** button the corresponding evaluation survey is displayed, Fig. 3.16.

The data in table are highlighted according to the red-yellow-green scale which is ranging from red over yellow to green. Green means goal has been achieved, yellow corresponds an average evaluation, red means goal has been failed. Each row from 1 until 15 contains a checklist data record. The dominating green values are a first signal that the course is running well.

Right last column with head $Q_e^{*(k)}(C)$ indicates empirical evaluation scores for obtained checklist data records for $k = 1, \ldots, 15$ records. Row with $Q^*(A_{ij})$ shows empirical scores of each sub goals. These scores are identical with the empirical evaluation scores $Q_e^*(A_{ij})$. Last row with $Q_e^*(B_i)$ shows the empirical key goal evaluation scores. Finally, on right end of bottom line the empirical evaluation score $Q_e^*(C)$ of the whole sample, the main result of data analysis is displayed.

Online tool for data processing generates moreover additional information with regard to color scale and confidence intervals for empirical evaluation scores, see

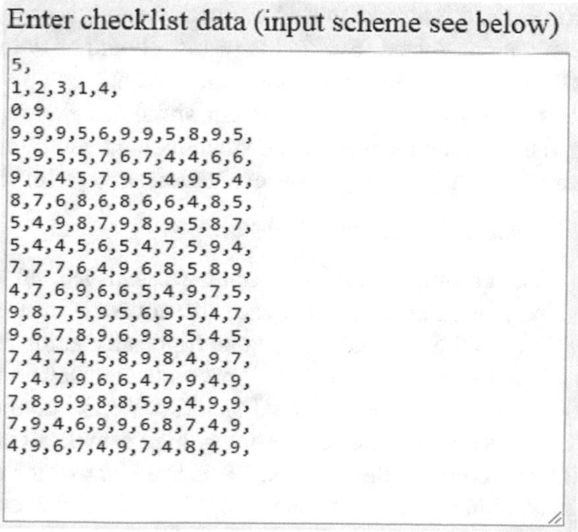

Fig. 3.15 Command lines in CSV format in online calculation window, satisfied students

Enter checklist data (input scheme see below)

```
5,
1,2,3,1,4,
0,9,
9,9,9,5,6,9,9,5,8,9,5,
5,9,5,5,7,6,7,4,4,6,6,
9,7,4,5,7,9,5,4,9,5,4,
8,7,6,8,6,8,6,6,4,8,5,
5,4,9,8,7,9,8,9,5,8,7,
5,4,4,5,6,5,4,7,5,9,4,
7,7,7,6,4,9,6,8,5,8,9,
4,7,6,9,6,6,5,4,9,7,5,
9,8,7,5,9,5,6,9,5,4,7,
9,6,7,8,9,6,9,8,5,4,5,
7,4,7,4,5,8,9,8,4,9,7,
7,4,7,9,6,6,4,7,9,4,9,
7,8,9,9,8,8,5,9,4,9,9,
7,9,4,6,9,9,6,8,7,4,9,
4,9,6,7,4,9,7,4,8,4,9,
```

	B_1	B_2		B_3			B_4	B_5				
k	A_{11}	A_{21}	A_{22}	A_{31}	A_{32}	A_{33}	A_{41}	A_{51}	A_{52}	A_{53}	A_{54}	$Q^*_{e,k}(C)$
1	9	9	9	5	6	9	9	5	8	9	5	1
2	5	9	5	5	7	6	7	4	4	6	6	0.7
3	9	7	4	5	7	9	5	4	9	5	4	0.82
4	8	7	6	8	6	8	6	6	4	8	5	0.76
5	5	4	9	8	7	9	8	9	5	8	7	0.87
6	5	4	4	5	6	5	4	7	5	9	4	0.58
7	7	7	7	6	4	9	6	8	5	8	9	0.83
8	4	7	6	9	6	6	5	4	9	7	5	0.71
9	9	8	7	5	9	5	6	9	5	4	7	0.89
10	9	6	7	8	9	6	9	8	5	4	5	0.87
11	7	4	7	4	5	8	9	8	4	9	7	0.81
12	7	4	7	9	6	6	4	7	9	4	9	0.74
13	7	8	9	9	8	8	5	9	4	9	9	0.85
14	7	9	4	6	9	9	6	8	7	4	9	0.88
15	4	9	6	7	4	9	7	4	8	4	9	0.81
$Q^*(A_{ij})$	0.76	0.76	0.72	0.73	0.73	0.83	0.71	0.74	0.67	0.73	0.74	$Q^*_e(C)=0.807$
$Q^*_e(B_i)$	0.76	0.81		0.92			0.71	0.93				

Fig. 3.16 Table with evaluation scores for satisfied students

Fig. 3.17. This information helps the evaluator to summarize the results for the groups involved.

In analogy, corresponding overviews are obtained for the unsatisfied and gambling students, see Figs. 3.18 and 3.19.

Step 8. Evaluation report

Report and interpretation of evaluation result is one of most important steps of any evaluation process. Evaluation results should be delivered in an appropriate time to all involved groups with same explanation.

In this example three different simulated evaluation data sets are investigated: satisfied students, unsatisfied students, and gambling students. Figures 3.16, 3.17, and 3.18, respectively, show evaluation surveys for these three cases.

Example 3.2.1 The satisfied students.

The empirical evaluation score for satisfied students is $Q^*_e(C) = 0.807$. This value is greater than 0.5. Hence the course is evaluated above average. The value $Q^*_e(C) = 0.807$ stands for a quite high satisfaction of students with course.

The lower table of Fig. 3.12 presents confidence intervals for evaluation score at the confidence levels 0.90, 0.95, and 0.99. The confidence intervals are quite wide. That is a consequence of the not so high sample size of $n = 15$ here. No confidence interval contains the value 0.5. This is a hint that the deviation of obtained empirical evaluation score 0.80 from value 0.5 can be considered as significant.

In summary the sample result indicates a significant above-average satisfaction of students with course in an average rate of approximately 81%.

Color scale

0	0.9	1.8	2.7	3.6	4.5	5.4	6.3	7.2	8.1	9
0	0.1	0.2	0.3	0.4	0.5	0.6	0.7	0.8	0.9	1

Empirical evaluation score $Q_e^*(C)$

$$Q_e^*(C) = 0.807000$$

Asymptotic confidence intervals $[q_{e,0}^*, q_{e,1}^*]$ for $Q_e(C)$ at confidence level
$1 - \alpha = 0.90, 0.95, 0.99$
and sample standard deviation σ_e^* with $\sigma_e^* = \sqrt{\frac{1}{n-1} \sum_{k=1}^{n} (Q_{e,k}^*(C) - Q_e^*(C))^2}$.

$1 - \alpha$	$q_{e,0}^*$	$Q_e^*(C)$	$q_{e,1}^*$	σ_e^*
0.90	0.7648		0.8492	
0.95	0.7567	0.807	0.8573	0.0994
0.99	0.7409		0.8731	

Sample size $n = 15$

Fig. 3.17 Additional information from online tool for satisfied students

The empirical evaluation scores for key goals have values near the empirical evaluation score for total goal. That means, the achievement of key goals is evaluated in the same direction as the achievement of total course goal. There is no evidence for particular weak points in view of the key goals to be reached.

Example 3.2.2 The unsatisfied students.

The students are unsatisfied in broad front with course. The general score set is again the set $\{0, 1, \ldots, 9\}$. The students evaluate the success of single goals between 0 and 5. Table in Fig. 3.18 shows the checklist results. The scores are now realizations of a uniformly distributed random variable with values in the set $\{0, 1, \ldots, 5\}$. The dominating colors are red, orange until yellow. This is a first hint that something might have went wrong with that course.

The values for the empirical evaluation scores of key goals $B_1, \ldots, B_5$ are quite high yet. They range from 0.24 until 0.35, but the empirical evaluation score $Q_e^*(C)$ is only 0.17. This is a comparatively small value and seems to be a contradiction to the values of empirical scores for key goals. In fact this value shows that the SURE model is working correctly. The SURE model includes the whole logical structure of total goal into evaluation. This goal is failed if only one of key goals has been failed. This is finally the reason for an empirical evaluation score $Q_e^*(C)$ of only 0.17.

k	B_1	B_2		B_3			B_4	B_5				$Q_{e,k}^*(C)$
	A_{11}	A_{21}	A_{22}	A_{31}	A_{32}	A_{33}	A_{41}	A_{51}	A_{52}	A_{53}	A_{54}	
1	5	4	3	5	5	4	4	1	0	5	5	0.45
2	4	0	5	3	1	5	4	5	4	3	4	0.4
3	1	3	2	1	2	3	0	4	5	0	2	0
4	3	0	2	1	4	3	2	4	3	1	4	0.25
5	2	4	0	1	3	5	0	0	3	1	3	0
6	0	3	0	3	4	5	3	5	1	0	5	0
7	4	2	4	1	5	2	5	0	0	1	4	0.34
8	1	3	1	0	4	2	4	0	5	4	4	0.26
9	0	1	2	5	5	2	1	5	1	5	2	0
10	5	1	1	2	0	1	3	2	0	4	0	0.21
11	2	4	5	4	2	5	4	1	3	4	2	0.36
12	0	4	1	2	5	1	4	5	4	4	0	0
13	0	0	1	2	4	4	3	2	4	5	1	0
14	3	3	3	3	2	4	0	1	1	0	5	0
15	3	3	5	5	2	4	4	0	5	5	1	0.4
$Q^*(A_{ij})$	0.24	0.26	0.26	0.28	0.36	0.37	0.3	0.26	0.29	0.31	0.31	$Q_e^*(C) = 0.1776$
$Q_e^*(B_i)$	0.24	0.27		0.35			0.3	0.32				

Fig. 3.18 Evaluation scores for unsatisfied students

This behavior of SURE model is an essential advantage against linear evaluation models. Linear models react in such cases quite insensible. They are not able to recognize such situations.

Example 3.2.3 The gambling students.
This is a fictive situation where we assume that the students do not answer to the questions of checklist but select randomly a score value from $\{0, 1, \ldots, 9\}$ as answer.

Table in Fig. 3.19 contains a corresponding sampling result. No red or green colors are dominating now, all values between red and green are represented quite uniformly. The empirical evaluation scores of single goals range between 0.44 and 0.64 and reflect at first glance in tendency an average satisfaction level. The same applies to key goals. These scores range between 0.46 and 0.64. Nevertheless, the total empirical evaluation score $Q_e^*(C)$ amounts only to 0.39 which is quite far away from an average satisfaction at a rate of 50%. Again this is a hint to the right direction. If we consider the scores $Q_e^{*(k)}(C)$, $k = 1, \ldots, 15$, of checklist data records, then it appears that the total goal has been failed for four times.

Hence the examples underline as well the advantages as the necessity of the structure oriented evaluation. Non-observance of logical goal structure may cause misjudgment.

The wide confidence intervals as well as the comparatively high value of empirical standard deviation $\sigma_e = 0.27$ are a hint to a particularly pronounced random background during interrogation.

k	B_1	B_2		B_3			B_4	B_5				$Q^*_{e,k}(C)$
	A_{11}	A_{21}	A_{22}	A_{31}	A_{32}	A_{33}	A_{41}	A_{51}	A_{52}	A_{53}	A_{54}	
1	5	9	7	9	9	4	5	2	2	8	9	0.79
2	6	0	0	7	3	3	5	2	6	2	8	0
3	3	6	9	1	6	2	5	1	2	0	1	0.38
4	9	9	0	4	5	3	7	3	2	3	3	0.64
5	3	0	7	6	6	9	2	8	8	6	5	0.5
6	7	8	1	0	3	3	3	3	4	2	3	0.43
7	5	7	8	8	4	2	1	8	6	4	8	0.48
8	3	1	5	3	3	3	1	0	8	9	9	0.34
9	0	9	2	3	2	6	7	5	0	2	3	0
10	8	0	8	1	1	6	7	6	5	1	7	0.63
11	0	5	0	4	6	1	3	6	0	3	6	0
12	3	6	8	0	3	4	0	5	8	9	6	0
13	6	0	7	3	8	3	8	9	1	9	5	0.72
14	2	3	1	3	7	8	4	0	4	8	6	0.4
15	2	8	0	7	8	6	6	8	3	2	8	0.56
$Q^*(A_{ij})$	0.46	0.53	0.47	0.44	0.55	0.47	0.47	0.49	0.44	0.5	0.64	$Q^*_e(C) = 0.3919$
$Q^*_e(B_i)$	0.46	0.64		0.55			0.47	0.64				

Fig. 3.19 Evaluation scores for gambling students

3.3 Evaluation of Active Learning Method in Flipped Class

The Chemnitz University of Technology (CUT) takes part in the international project DrIVE-MATH funded by Erasmus Plus. This project runs from May 2017 to July 2020. The Faculty of Computer Science at CUT is responsible for the project. In frame of the project, the Computer Engineering Professorship has to implement some of the active learning methods into engineering education at CUT. Therefore, the Computer Engineering Professorship implemented eduScrum, Jigsaw, Problem based learning, and Hands-on methods as active learning methods into their teaching.

To recognize efficiency of implemented methods, the project team conducted a formative evaluation for eduScrum active learning method. The eduScrum test class consisted of volunteer students of the master program Automotive Software Engineering (ASE) at CUT. The new teaching approach eduScrum was applied during the winter semester 2018/19 and the summer semester 2019. The teacher of the subject invited the students to take part at the eduScrum test class. In the first round 30 of 80 students registered to the eduScrum class, during the second round 40 of 85 students selected the eduScrum test class (Tudevdagva et al. 2020).

Evaluation goal describes the objectives of the evaluation process, which should be reached successfully after the implementation of the learning method.

The five main objectives also named key goals are defined as follow (Fig. 3.20):

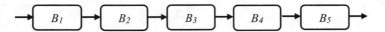

Fig. 3.20 The key goal structure for evaluation of active learning method

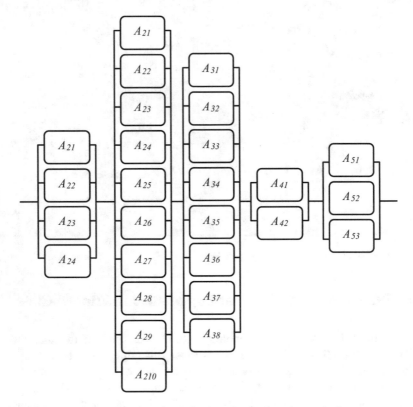

Fig. 3.21 The sub key goal structure for evaluation of active learning method

- B_1: Students' acceptance of the eduScrum method;
- B_2: Advantages of the active learning method compared to traditional teaching;
- B_3: Improvement of the soft skills of the students while using active learning, namely eduScrum;
- B_4: The knowledge acquired in theory and practice while using the active learning;
- B_5: Importance and essential of active learning while teaching engineering subjects.

Step 2. Definition of sub goals.

Next step of the SURE evaluation is the definition of sub goals based on the key goal structure (Fig. 3.21).

The first key goal B_1 consists of four sub goals A_{11}, A_{12}, A_{13}, and A_{14}. Therefore, four questions are focused on the feedback from the students on their opinion and acceptance about eduScrum as a teaching approach.

The second key goal B_2 consists of ten sub goals $A_{21}, A_{22}, A_{23}, A_{24}, A_{25}, A_{26}, A_{27}, A_{28}, A_{29}$, and A_{210}. All ten questions are directed to collect students' opinions with regard to the comparison of active learning method eduScrum with traditional teaching.

The third key goal B_3 consists of eight sub goals $A_{31}, A_{32}, A_{33}, A_{34}, A_{35}, A_{36}, A_{37}$, and A_{38}. These sub goals are focused on the measurement of the impact of eduScrum to increase the soft skills of the participants.

The fourth key goal B_4 consists of two sub goals A_{41} and A_{42}. These two criteria are targeted to measure advantages of eduScrum active learning method compared to traditional teaching methods.

The fifth key goal B_5 consists of three sub goals A_{51}, A_{52}, and A_{53}. Via these three questions a feedback from students about future implementation of AL methods in engineering courses is wanted.

Step 3. Confirmation of goal structures

The results of the previous two steps are logical structure of evaluation goals. Those structures should be checked by involved groups and confirmed by an official protocol. This step has been managed by online discussions exchange of emails from partner universities.

Step 4. Creation of checklist

This step is one of the most important steps for any kind of evaluation process. Four universities from Germany, France, Slovakia, and Portugal were involved in this evaluation. According to the project plan the teams met several times and discussed the checklist. After meeting of June 2019 in Chemnitz, a draft of a checklist for evaluation was completed. All universities worked on the draft from distance and the updated versions could be seen by all partners via an online box. After several updates checklist for evaluation was ready.

All checklist elements are evaluated according the following scale:

- 1 = Strongly Disagree;
- 2 = Disagree;
- 3 = Neither agree nor disagree (Neutral);
- 4 = Agree;
- 5 = Strongly Agree.

The first key goal B_1, formulated as "Students acceptance of the eduScrum method," has four sub goals. Criteria for those sub goals are formulated as shown in Fig. 3.22.

The second key goal B_2, formulated as "Advantages of the active learning method compared to traditional teaching," has ten sub goals. Criteria for those sub goals are formulated as shown in Fig. 3.23.

The third key goal B_3, formulated as "Improvement of the soft skills of the students while using active learning, namely eduScrum," has eight sub goals. Criteria for those sub goals are formulated as shown in Fig. 3.24.

S/N	Criteria	Strongly Disagree	Disagree	Neutral	Agree	Strongly Agree
1.	The active learning is an useful learning strategy					
2.	The use of innovative active-learning methods should be an integral part of engineering education					
3.	I can apply learned skills into other scenarios					
4.	I would recommend my colleagues to use it for study purposes.					

Fig. 3.22 Questions of sub goals A_{11} to A_{14}

Compared to traditional teaching approach the active-learning method:

S/N	Criteria	Strongly Disagree	Disagree	Neutral	Agree	Strongly Agree
1.	... increases my interest in learning					
2.	...increases my motivation in studying					
3.	...motivates me to study more effectively					
4.	...encourage me to study more deeply					
5.	... are more demanding my preparation					
6.	... are more beneficial for gaining knowledge					
7.	... enlarges my professional insight					
8.	... are more creative					
9.	... are more convenient and pleasant					
10.	... are more accessible and comfortable					

Fig. 3.23 Questions for sub goals from A_{21} to A_{210}

The fourth key goal B_4, formulated as "The knowledge acquired in theory and practice while using active learning," has two sub goals. Criteria for those sub goals are formulated as shown in Fig. 3.25.

The fifth key goal B_5, formulated as "Importance and essential of active learning while teaching engineering subjects," has three sub goals. Criteria for those sub goals are formulated as shown in Fig. 3.26.

Figures 3.23, 3.24, 3.25, and 3.26 show adapted checklists for data collections. The checklists were created based on defined sub goals.

Step 5. Confirmation of checklist

When the adapted checklist is ready, the next step can start. This is the confirmation step of checklist. The created checklist was located in a cloud folder to which the involved universities had access to. The developed checklist was accepted by all involved universities and confirmed by an official protocol.

Step 6. Data collection

The following soft skills were developed by active learning:

S/N	Criteria	Strongly Disagree	Disagree	Neutral	Agree	Strongly Agree
1.	Leadership					
2.	Personal and social responsibility					
3.	Flexibility/adaptability					
4.	Interest in teamwork					
5.	Negotiation and conflict resolution					
6.	Professionalism/ethics					
7.	Empathic behavior					
8.	Creativity					

Fig. 3.24 Questions of sub goals from A_{31} to A_{38}

The knowledge acquired by active-learning methods:

S/N	Criteria	Strongly Disagree	Disagree	Neutral	Agree	Strongly Agree
1.	integrates theory and practice					
2.	being more applicable in my professional orientation					

Fig. 3.25 Questions of sub goals A_{41} and A_{42}

Active learning methods in engineering education

S/N	Criteria	Strongly Disagree	Disagree	Neutral	Agree	Strongly Agree
1.	The use of innovative active-learning methods should be an integral part of engineering education					
2.	Active learning methods are not always needed to use in engineering education					
3.	EduScrum is excellent active learning method for engineering education					

Fig. 3.26 Questions of sub goals A_{51}, A_{52}, and A_{53}

For data collection CUT team created an online questionnaire. The online questionnaire was developed using the ONYX tool. ONYX Testsuite is a complex test system based on the IMS-QTI-2.1 standard, which was developed in cooperation with the Saxon universities by BPS Bildungsportal Sachsen GmbH. This is the so-called All in one: self tests, tests, exams, and questionnaire tool for online assessment.

Figure 3.27 shows a screenshot of online questionnaire in OPAL learning management system. This link was sent to all students, who participated in

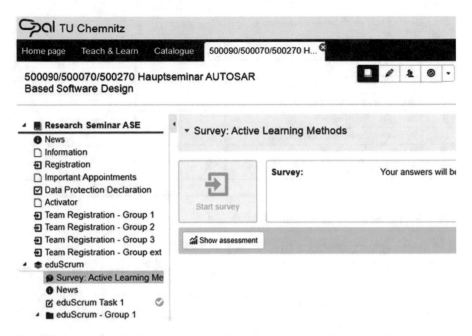

Fig. 3.27 Screenshot of online questionnaire for evaluation of active learning method

eduScrum class. Around 23% of the eduScrum students sent a reply to the requested questionnaire assessment.

In total 15 valid answers were received and used for data processing.

$$5, 5, 5, 5, 5, 5, 5, 5, 5, 5, 5, 5, 5, 5, 5, 5, 5, 5, 5, 5, 5, 5, 5, 5, 5, 5, 5, 5, 5,$$

$$4, 5, 4, 4, 4, 4, 4, 4, 4, 4, 4, 4, 4, 4, 4, 4, 4, 4, 4, 4, 4, 4, 4, 4, 4, 4, 4, 4,$$

$$4, 4, 3, 3, 3, 3, 3, 3, 3, 3, 3, 3, 3, 3, 2, 2, 2, 2, 2, 2, 2, 2, 3, 3, 3, 3, 2,$$

$$5, 5, 5, 5, 5, 5, 5, 5, 5, 5, 5, 5, 5, 5, 5, 4, 5, 5, 5, 5, 5, 5, 5, 5, 5, 5, 5,$$

$$4, 4, 3, 4, 3, 3, 3, 4, 3, 3, 3, 3, 4, 3, 2, 2, 3, 2, 2, 3, 3, 3, 4, 4, 3, 3, 3,$$

$$4, 4, 3, 4, 4, 4, 4, 4, 3, 4, 3, 3, 4, 4, 3, 3, 4, 3, 3, 3, 3, 3, 4, 4, 4, 4, 3,$$

$$5, 5, 5, 5, 5, 5, 5, 5, 5, 5, 5, 5, 5, 5, 5, 5, 5, 5, 5, 5, 5, 5, 5, 5, 5, 5, 5,$$

$$4, 4, 3, 4, 4, 4, 4, 4, 4, 4, 4, 3, 4, 4, 3, 4, 4, 3, 3, 3, 3, 4, 4, 4, 4, 4, 3,$$

$$5, 5, 5, 5, 5, 5, 5, 5, 5, 5, 5, 5, 5, 5, 5, 5, 5, 5, 5, 5, 5, 5, 5, 5, 5, 5, 5,$$

$$4, 4, 3, 4, 4, 4, 4, 4, 4, 4, 4, 4, 4, 4, 3, 4, 4, 4, 4, 4, 3, 4, 4, 4, 4, 4, 4,$$

$$5, 5, 5, 5, 5, 5, 5, 5, 4, 4, 5, 5, 5, 5, 4, 4, 4, 4, 4, 5, 4, 5, 4, 5, 5, 4, 5,$$

$$4, 5, 4, 4, 4, 4, 4, 4, 4, 4, 4, 4, 4, 4, 4, 4, 4, 4, 4, 4, 3, 4, 4, 4, 4, 4, 4,$$

5, 5, 4, 4, 4, 4, 4, 4, 4, 4, 4, 5, 4, 4, 4, 4, 4, 4, 4, 4, 4, 4, 4, 4, 4, 4, 4,

5, 5, 5, 5, 4, 4, 5, 4, 4, 4, 5, 5, 5, 5, 4, 4, 4, 4, 4, 4, 4, 4, 4, 4, 4, 4, 4,

5, 5, 5, 5, 5, 5, 5, 5, 4, 5, 5, 5, 5, 5, 4, 4, 5, 4, 4, 5, 5, 5, 4, 5, 5, 4, 5,

Step 7. Data processing

For data processing, the online tool of the SURE model was used.

As explained in Example 3.2, the collected data were put in CSV format into input window of the tool. The following options were selected: red-yellow-green color scale type; non-normalized checklist data display format and empirical evaluation scores for evaluation table. After clicking on the button **SEND** the evaluation scores of Fig. 3.28 were obtained.

By the online tool of the SURE model, data processing is very convenient. One only needs to enter collected data in correct format and press the **SEND** button. After few moments the evaluation scores will be displayed.

Step 8. Evaluation report

Empirical evaluation scores were calculated by online tool of the SURE model. All empirical scores for sub goals are equal or higher than 0.71. It can be summarized that the 15 students from the eduScrum class have evaluated the implementation of the active learning method eduScrum very positive. However, this result reflects a trend and evaluation cannot be used to interpret the feedback of the remaining part of students but with a tendency it confirms a successful eduScrum implementation.

Figure 3.28 shows that 3 of 15 students evaluated always all criteria with very satisfied. Next 6 students of the 15 students answered only very satisfied and satisfied. With respect of the somewhat low number of replies of students the observed data cannot be a result of all students. The data show that for the next round the teachers should direct more attention on motivating the students to participate in the survey.

This time the evaluation questions were sent to the students after the course during the exam period. So the teacher summarized that this could be one of reasons for the low participation in the evaluation questionnaire.

The highest evaluation score of 0.88 was obtained for the question "Active learning is a useful learning strategy." This score is motivation for the project team to continue with eduScrum as the method in the next round. This fits to expected result of the project investors, too. Therefore, the active learning method eduScrum will be implemented in the coming winter semester 2019/20 at CUT.

3.4   The Evaluation of Projects for Investment Selection

This evaluation was an order from the Association for the Development of Mongolian Women in Europe. The Association for the Development of Mongolian Women in Europe was established in November 2011. The Association for the Mongolian

k	B_1				B_2									
	A_{11}	A_{12}	A_{13}	A_{14}	A_{21}	A_{22}	A_{23}	A_{24}	A_{25}	A_{26}	A_{27}	A_{28}	A_{29}	A_{210}
1	5	5	5	5	5	5	5	5	5	5	5	5	5	5
2	4	5	4	4	4	4	4	4	4	4	4	4	4	4
3	4	4	3	3	3	3	3	3	3	3	3	3	3	3
4	5	5	5	5	5	5	5	5	5	5	5	5	5	5
5	4	4	3	4	3	3	3	4	3	3	3	3	4	3
6	4	4	3	4	4	4	4	4	3	4	3	3	4	4
7	5	5	5	5	5	5	5	5	5	5	5	5	5	5
8	4	4	3	4	4	4	4	4	4	4	4	3	4	4
9	5	5	5	5	5	5	5	5	5	5	5	5	5	5
10	4	4	3	4	4	4	4	4	4	4	4	4	4	4
11	5	5	5	5	5	5	5	5	4	5	5	5	5	5
12	4	5	4	4	4	4	4	4	4	4	4	4	4	4
13	5	5	4	4	4	4	4	4	4	4	4	5	4	4
14	5	5	5	5	4	4	5	4	4	4	5	5	5	5
15	5	5	5	5	5	5	5	5	4	5	5	5	5	5
$Q^*(A_{ij})$	0.88	0.92	0.78	0.85	0.82	0.82	0.83	0.83	0.77	0.8	0.82	0.82	0.85	0.83
$Q_e^*(B_i)$	0.9				0.85									

k	B_3								B_4		B_5			$Q_{e,k}^*(C)$
	A_{31}	A_{32}	A_{33}	A_{34}	A_{35}	A_{36}	A_{37}	A_{38}	A_{41}	A_{42}	A_{51}	A_{52}	A_{53}	
1	5	5	5	5	5	5	5	5	5	5	5	5	5	1
2	4	.4	4	4	4	4	4	4	4	4	4	4	4	0.79
3	2	2	2	2	2	2	2	2	3	3	3	3	2	0.44
4	5	4	5	5	5	5	5	5	5	5	5	5	5	1
5	2	2	3	2	2	3	3	3	4	4	3	3	3	0.57
6	3	3	4	3	3	3	3	3	4	4	4	4	3	0.67
7	5	5	5	5	5	5	5	5	5	5	5	5	5	1
8	3	4	4	3	3	3	3	4	4	4	4	4	3	0.7
9	5	5	5	5	5	5	5	5	5	5	5	5	5	1
10	3	4	4	4	4	4	3	4	4	4	4	4	4	0.73
11	4	4	4	4	4	5	4	5	5	4	5	4	5	1
12	4	4	4	4	4	4	3	4	4	4	4	4	4	0.79
13	4	4	4	4	4	4	4	4	4	4	4	4	4	0.84
14	4	4	4	4	4	4	4	4	4	4	4	4	4	0.84
15	5	4	5	4	4	5	5	5	4	5	5	4	5	1
$Q^*(A_{ij})$	0.72	0.72	0.78	0.72	0.72	0.77	0.72	0.78	0.8	0.83	0.82	0.78	0.77	$Q_e^*(C) = 0.8248$
$Q_e^*(B_i)$	0.76								0.83		0.8			

Fig. 3.28 Empirical evaluation scores, active learning method in flipped class

Women's Development in Europe (ADMWE) which is a non-profit organization was established to facilitate the Mongolian women's networking and development in Europe.

The Association for the Mongolian women's development in Europe's aim is to bring together the women of different professions and backgrounds and to provide opportunities to network and develop business and social connections, making ADMWE the ultimate networking circle.

The main roles of ADMWE are:

- To facilitate women's networking in Europe.
- To empower women at all levels of governmental, economic, and social life
- To promote the interests of the Mongolian women and help them to succeed in business and in life.
- To enhance recognition of women's success in the society.

Each year the ADMWE funds a project with different focus. Main challenge of the association is to select the most fitting project for a fund focus. This year five projects applied for a fund amounting 3000€. By the decision of board members of association, this year the SURE model was applied for project evaluation. Aim of evaluation is to support decision making of board members to select best fitting project for funding purposes.

Step 1. Definition of key goals

Main property of the SURE model is design evaluation goal structure. In first step of the SURE model the evaluator has to define key goals of evaluation (Fig. 3.29).

The first step in applying the SURE model is always to design an adapted logical structure of the evaluation objective. To do this, the evaluator must first define the key goals of the evaluation (Fig. 3.29). In the discussion with the board members of ADMWE, five key goals were selected for the evaluation (Fig. 3.29):

- B_1: Project target group should be girls and women;
- B_2: Candidate institutions have to provide proof about enough human resource for project
- B_3: Opportunity to continue the idea of project after implementation
- B_4: Project implementation should be visible in national level
- B_5: Candidate institutions should have 2 or 3 years experience at least

Step 2. Definition of sub goals

Next step of the SURE evaluation is the definition of sub goals based on the key goal structure (Fig. 3.30).

The first key goal B_1 consists of two sub goals A_{11} and A_{12}. Here two questions are focused on the project target group.

The second key goal B_2 consists of four sub goals A_{21}, A_{22}, A_{23}, and A_{24}. All four questions are directed to figure out the capacity of human resource of candidate institutions.

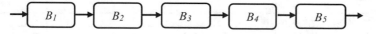

Fig. 3.29 The structure of key goals, evaluation of projects

The third key goal B_3 consists of two sub goals A_{31} and A_{32}. These sub goals focus on the opportunity to keep continuous development of project idea after implementation.

The fourth key goal B_4 consists of three sub goals A_{41}, A_{42}, and A_{43}. These three criteria are targeted to measure implementation level.

The fifth key goal B_5 consists of one sub goals A_{51}. A goal aimed to inspect the experience of candidate institutions.

Step 3. Confirmation of goal structures

Defined key and sub goal structures are sent to board members of association and confirmed by their acceptance.

Step 4. Creation of checklist

Checklist developed based on the sub goal structures. All questions are measured on the following scale:

All items are measured on the following scale:

- 0 = Strongly Disagree;
- 1 = Disagree;
- 2 = Neither agree nor disagree (Neutral);
- 3 = Agree;
- 4 = Strongly Agree.

The first key goal B_1 formulated as "Project target group should be girls and women" has two sub goals. Criteria for those sub goals are formulated as shown in Fig. 3.31.

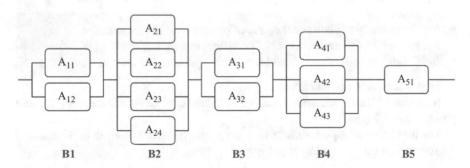

Fig. 3.30 The structure of sub goals, evaluation of projects

Project target group should be girls and women;

S/N	Criteria	Strongly Disagree	Disagree	Neutral	Agree	Strongly Agree
1.	Fitting fully to this criteria					
2.	Proofed activities will focus to target groups					

Fig. 3.31 Questions of sub goals A_{11}, and A_{12} for evaluation of projects

Candidate institution have provide proof about enough human resource for project

S/N	Criteria	Strongly Disagree	Disagree	Neutral	Agree	Strongly Agree
1.	Proofed capacity if human resource for project implementation					
2.	Not included not relevant cost into project plan					
3.	Not included institutions' yearly cost into project plan					
4.	Equipment cost always will be less than one million T					

Fig. 3.32 Questions for sub goals A_{21}, A_{22}, A_{23}, and A_{24} for evaluation of projects

Opportunity to continue idea of project after implementation

S/N	Criteria	Strongly Disagree	Disagree	Neutral	Agree	Strongly Agree
1.	Proofed that can continue idea of project after implementation					
2.	There are enough proof about sustainable activities					

Fig. 3.33 Questions for sub goals A_{31} and A_{32} for evaluation of projects

Project implementation should be visible in national level

S/N	Criteria	Strongly Disagree	Disagree	Neutral	Agree	Strongly Agree
1.	Institution can implement project in national level					
2.	Developed plan for visibility of project implementation in national level					
3.	In plan included several ways of project implementation					

Fig. 3.34 Questions for sub goals A_{41}, A_{42}, and A_{43} for evaluation of projects

The second key goal B_2 formulated as "Candidate institutions have to provide proof about enough human resource for project" has four sub goals. Criteria for those sub goals are formulated as shown in Fig. 3.32.

The third key goal B_3 formulated as "Opportunity to continue the idea of project after implementation" has two sub goals. Criteria for those sub goals are formulated as shown in Fig. 3.33.

The fourth key goal B_4 formulated as "Project implementation should be visible in national level" has three sub goals. Criteria for those sub goals are formulated as shown in Fig. 3.34.

The fifth key goal B_5 formulated as "Candidate institutions should have 2 or 3 years experience at least" has one sub goals. Criteria for those sub goals are formulated as shown in Fig. 3.35.

Step 5. Created checklist sent to board members of ADMWE. Every board member have to check questions and send back to evaluator confirmation as

Candidate institution should have from 2 or 3 years experience at least

S/N	Criteria	Strongly Disagree	Disagree	Neutral	Agree	Strongly Agree
1.	This institution works in giving field more than 2 years					

Fig. 3.35 Questions for sub goal A_{51} for evaluation of projects

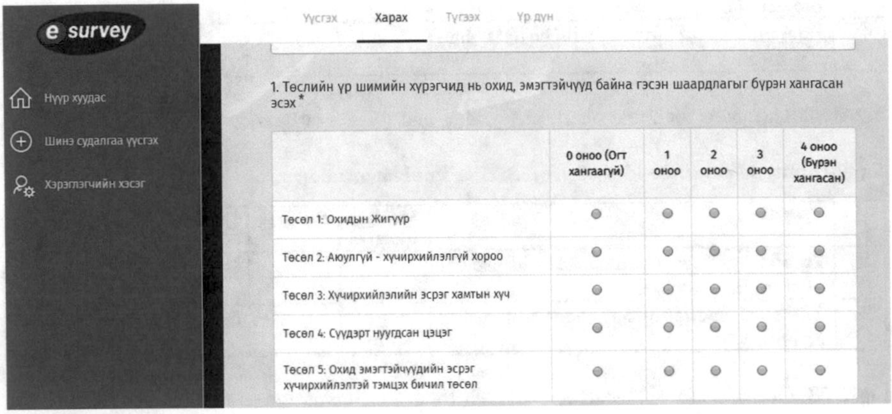

Fig. 3.36 Main page of eSurvey free software

accepted the checklist. After two weeks all confirmations received and the checklist proofed by board of ADMWE. Only proofed checklist can use for evaluation process.

Step 6. Data collection

Data was collected with online tool: http://app.esurvey.mn/ (Fig. 3.36). This is a free online tool for creation of checklists and questionnaires.

The board of association consists of 24 persons. Link of the online tool sent to all board members. In given time only 18 of the 24 board members reacted to online evaluation. After checking of the collected data only 10 of the responses were valid.

Each participant had to evaluate five projects by same questions.

Data for Project 1:

$$1, 4, 4, 0, 3, 4, 1, 1, 1, 0, 0, 1,$$
$$1, 1, 0, 0, 0, 1, 0, 1, 1, 1, 0, 2,$$
$$4, 4, 0, 0, 4, 4, 2, 0, 0, 0, 2, 0,$$
$$2, 1, 2, 0, 4, 2, 1, 1, 0, 1, 1, 2,$$
$$4, 4, 2, 0, 4, 4, 1, 1, 0, 1, 1, 2,$$
$$4, 3, 1, 4, 4, 2, 2, 1, 1, 1, 1, 4,$$
$$1, 2, 1, 0, 0, 4, 1, 2, 0, 0, 0, 1,$$

2, 2, 2, 2, 2, 2, 2, 2, 2, 2, 2, 2,

1, 1, 1, 1, 0, 4, 0, 0, 1, 1, 1, 2,

4, 4, 0, 0, 4, 4, 1, 1, 3, 3, 3, 1

Data for Project 2:

4, 4, 2, 1, 3, 4, 2, 2, 2, 2, 1, 1,

3, 1, 0, 1, 1, 1, 1, 2, 2, 1, 3, 1,

4, 4, 0, 0, 4, 4, 2, 0, 0, 0, 2, 0,

3, 3, 4, 3, 4, 4, 2, 2, 3, 2, 2, 3,

4, 4, 2, 1, 4, 4, 1, 1, 0, 1, 1, 2,

4, 4, 1, 4, 4, 2, 2, 1, 1, 1, 1, 4,

2, 2, 2, 4, 4, 4, 4, 4, 0, 0, 0, 1,

1, 1, 1, 1, 1, 1, 2, 2, 2, 2, 1, 2,

2, 1, 0, 0, 2, 0, 0, 0, 3, 3, 3, 3,

4, 4, 1, 0, 1, 4, 2, 1, 3, 2, 2, 3

Data for Project 3:

3, 3, 4, 4, 4, 3, 3, 3, 2, 2, 1, 4,

4, 4, 4, 4, 4, 3, 4, 4, 4, 4, 4, 4,

4, 4, 4, 4, 4, 3, 3, 4, 2, 4, 2, 3,

4, 3, 2, 4, 4, 4, 1, 0, 0, 1, 1, 0,

4, 4, 4, 4, 4, 0, 1, 1, 0, 4, 3, 4,

4, 3, 1, 4, 4, 1, 2, 1, 1, 2, 3, 4,

4, 3, 3, 4, 0, 0, 3, 1, 0, 1, 1, 2,

4, 4, 4, 4, 4, 4, 4, 4, 4, 4, 4, 4,

4, 4, 4, 4, 4, 4, 4, 4, 4, 4, 4, 4,

4, 4, 4, 4, 4, 1, 3, 4, 4, 4, 1, 4

Data for Project 4:

3, 1, 2, 4, 4, 4, 3, 3, 4, 4, 4, 4,

1, 2, 0, 2, 1, 1, 0, 1, 1, 1, 4, 2,

4, 4, 4, 4, 4, 4, 4, 4, 4, 4, 3, 4,

1, 1, 2, 4, 4, 4, 1, 0, 4, 1, 0, 3,

4, 4, 4, 4, 4, 4, 4, 4, 4, 4, 4, 4,

4, 2, 1, 4, 4, 2, 2, 1, 3, 2, 4, 4,

4, 3, 3, 4, 0, 4, 4, 3, 4, 2, 2, 2,

3, 3, 3, 3, 3, 3, 3, 3, 3, 3, 4, 3,

1, 1, 1, 1, 1, 4, 0, 1, 1, 1, 4, 3

4, 4, 1, 4, 1, 3, 2, 4, 4, 4, 4, 4,

Data for Project 5:

4, 2, 2, 4, 4, 4, 1, 1, 4, 4, 4, 2,

2, 1, 0, 0, 1, 1, 0, 1, 2, 1, 1, 1,

4, 4, 4, 4, 4, 4, 4, 4, 4, 4, 4, 2,

1, 1, 3, 4, 4, 4, 1, 1, 3, 2, 1, 1,

4, 2, 4, 4, 4, 4, 3, 1, 4, 3, 2, 0,

4, 3, 1, 4, 4, 2, 2, 1, 1, 1, 1, 4,

1, 1, 0, 0, 2, 4, 2, 2, 4, 0, 1, 0,

1, 1, 1, 1, 1, 1, 1, 1, 1, 1, 1, 1,

1, 0, 0, 0, 0, 0, 0, 0, 3, 3, 1, 1,

4, 3, 3, 4, 4, 4, 1, 4, 2, 4, 1, 3

Step 7. Data processing

For data processing online tool of the SURE model was used.

As explained in example 3.2 collected data must be entered into the online tool in CSV format. The following options have been selected: red-yellow-green color scale type; non-normalized checklist data display format and empirical evaluation scores for evaluation table.

Figures 3.37, 3.38, 3.39, 3.40, and 3.41 contain the corresponding evaluation results.

Step 8. Evaluation report

Five projects were compared and corresponding evaluation results were calculated. Table in Fig. 3.42 contains the determined empirical evaluation scores for projects.

After the evaluation has been carried out two projects, projects 3 and 4, whole empirical scores are quite height, remain in the shortlist.

k	B_1		B_2				B_3		B_4			B_5	$Q^*_{e,k}(C)$
	A_{11}	A_{12}	A_{21}	A_{22}	A_{23}	A_{24}	A_{31}	A_{32}	A_{41}	A_{42}	A_{43}	A_{51}	
1	1	4	4	0	3	4	1	1	1	0	0	1	0.36
2	1	1	0	0	0	1	0	1	1	1	0	2	0.18
3	4	4	0	0	4	4	2	0	0	0	2	0	0
4	2	1	2	0	4	2	1	1	0	1	1	2	0.38
5	4	4	2	0	4	4	1	1	0	1	1	2	0.47
6	4	3	1	4	4	2	2	1	1	1	1	4	0.63
7	1	2	1	0	0	4	1	2	0	0	0	1	0
8	2	2	2	2	2	2	2	2	2	2	2	2	0.5
9	1	1	1	1	0	4	0	0	1	1	1	2	0
10	4	4	0	0	4	4	1	1	3	3	3	1	0.51
$Q^*(A_{ij})$	0.6	0.65	0.33	0.18	0.63	0.78	0.28	0.25	0.23	0.25	0.28	0.43	$Q^*_c(C) = 0.3058$
$Q^*_c(B_i)$	0.68		0.86				0.27		0.26			0.43	

Fig. 3.37 Evaluation result for Project 1

k	B_1		B_2				B_3		B_4			B_5	$Q^*_{e,k}(C)$
	A_{11}	A_{12}	A_{21}	A_{22}	A_{23}	A_{24}	A_{31}	A_{32}	A_{41}	A_{42}	A_{43}	A_{51}	
1	4	4	2	1	3	4	2	2	2	2	1	1	0.56
2	3	1	0	1	1	1	1	2	2	1	3	1	0.36
3	4	4	0	0	4	4	2	0	0	0	2	0	0
4	3	3	4	3	4	4	2	2	3	2	2	3	0.7
5	4	4	2	1	4	4	1	1	0	1	1	2	0.47
6	4	4	1	4	4	2	2	1	1	1	1	4	0.63
7	2	2	2	4	4	4	4	4	0	0	0	1	0
8	1	1	1	1	1	1	2	2	2	2	1	2	0.37
9	2	1	0	0	2	0	0	0	3	3	3	3	0
10	4	4	1	0	1	4	2	1	3	2	2	3	0.71
$Q^*(A_{ij})$	0.78	0.7	0.33	0.38	0.7	0.7	0.45	0.38	0.4	0.35	0.4	0.5	$Q^*_c(C) = 0.3781$
$Q^*_c(B_i)$	0.75		0.76				0.42		0.4			0.5	

Fig. 3.38 Evaluation result for Project 2

3.5 Summary

In this section a measure theoretical approach for evaluation of logically structured processes is developed. Via measure spaces for the involved e-learning processes a product measure space is generated. This product space allows the calculation of scores for evaluation of logical structured processes. The product measure satisfies the same calculation rules like any normalized measure of general measure theory.

Depending on contribution of process components to achievement of a process goal logical series and parallel structures play an important part. For these goal structures corresponding score calculation rules are derived. By combination of these rules a consistent evaluation of complex goal structures becomes possible.

	B_1		B_2				B_3		B_4			B_5	
k	A_{11}	A_{12}	A_{21}	A_{22}	A_{23}	A_{24}	A_{31}	A_{32}	A_{41}	A_{42}	A_{43}	A_{51}	$Q^*_{e,k}(C)$
1	3	3	4	4	4	3	3	3	2	2	1	4	0.75
2	4	4	4	4	4	3	4	4	4	4	4	4	1
3	4	4	4	4	4	3	3	4	2	4	2	3	0.94
4	4	3	2	4	4	4	1	0	0	1	1	0	0
5	4	4	4	4	4	0	1	1	0	4	3	4	0.76
6	4	3	1	4	4	1	2	1	1	2	3	4	0.73
7	4	3	3	4	0	0	3	1	0	1	1	2	0.55
8	4	4	4	4	4	4	4	4	4	4	4	4	1
9	4	4	4	4	4	4	4	4	4	4	4	4	1
10	4	4	4	4	4	1	3	4	4	4	1	4	1
$Q^*(A_{ij})$	0.98	0.9	0.85	1	0.9	0.58	0.7	0.65	0.53	0.75	0.6	0.83	$Q^*_e(C)=0.7735$
$Q^*_e(B_i)$	0.98		1				0.71		0.73			0.83	

Fig. 3.39 Evaluation result for Project 3

	B_1		B_2				B_3		B_4			B_5	
k	A_{11}	A_{12}	A_{21}	A_{22}	A_{23}	A_{24}	A_{31}	A_{32}	A_{41}	A_{42}	A_{43}	A_{51}	$Q^*_{e,k}(C)$
1	3	1	2	4	4	4	3	3	4	4	4	4	0.84
2	1	2	0	2	1	1	0	1	1	1	4	2	0.37
3	4	4	4	4	4	4	4	4	4	4	3	4	1
4	1	1	2	4	4	4	1	0	4	1	0	3	0.48
5	4	4	4	4	4	4	4	4	4	4	4	4	1
6	4	2	1	4	4	2	2	1	3	2	4	4	0.83
7	4	3	3	4	0	4	4	3	4	2	2	2	0.87
8	3	3	3	3	3	3	3	3	3	3	4	3	0.79
9	1	1	1	1	1	4	0	1	1	1	4	3	0.48
10	4	4	1	4	1	3	2	4	4	4	4	4	1
$Q^*(A_{ij})$	0.73	0.63	0.53	0.85	0.65	0.83	0.58	0.6	0.8	0.65	0.83	0.83	$Q^*_e(C)=0.7664$
$Q^*_e(B_i)$	0.72		0.9				0.63		1			0.83	

Fig. 3.40 Evaluation result for Project 4

Based on the calculation rules an estimation method for scores is developed. By this method empirical scores can be calculated for sampling results obtained by means of adapted checklists.

For a comparison of scores for evaluation of different processes a score calibration is recommended. This calibration transforms a score value into an evaluation score value. This value enables a further interpretation of score values and is helpful at final analysis of sampling data for evaluation report. An estimation for the evaluation score based on checklist results can be obtained like at empirical score.

The precision of empirical scores or empirical evaluation scores can be estimated by confidence intervals.

k	B_1		B_2				B_3		B_4			B_5	$Q^*_{e,k}(C)$
	A_{11}	A_{12}	A_{21}	A_{22}	A_{23}	A_{24}	A_{31}	A_{32}	A_{41}	A_{42}	A_{43}	A_{51}	
1	4	2	2	4	4	4	1	1	4	4	4	2	0.66
2	2	1	0	0	1	1	0	1	2	1	1	1	0.23
3	4	4	4	4	4	4	4	4	4	4	4	2	0.87
4	1	1	3	4	4	4	1	1	3	2	1	1	0.39
5	4	2	4	4	4	4	3	1	4	3	2	0	0
6	4	3	1	4	4	2	2	1	1	1	1	4	0.63
7	1	1	0	0	2	4	2	2	4	0	1	0	0
8	1	1	1	1	1	1	1	1	1	1	1	1	0.25
9	1	0	0	0	0	0	0	0	3	3	1	1	0
10	4	3	3	4	4	4	1	4	2	4	1	3	0.91
$Q^*(A_{ij})$	0.65	0.45	0.45	0.63	0.7	0.7	0.38	0.4	0.7	0.58	0.43	0.38	$Q^*_e(C)=0.3964$
$Q^*_e(B_i)$	0.63		0.74				0.43		0.7			0.38	

Fig. 3.41 Evaluation result for Project 5

Fig. 3.42 Empirical evaluation scores for projects

	Evaluation score
Project 1	0.3058
Project 2	0.3781
Project 3	0.7735
Project 4	0.7664
Project 5	0.3964

Some simulation examples are considered which illustrate for selected evaluation situations the developed evaluation method and demonstrate the advantages of SURE model against linear evaluation models.

The presented evaluation method is by their embedding into the general measure theory quite universal applicable. Other areas of application are, for instance, the evaluation of efficiency of administrative or organizational processes and the evaluation of robustness of embedded systems during run-time (see, e.g., Heller, 2013).

References

Englisch, E., Hardt, W., Heller, A., Tudevdagva, U., Tonndort-Martini, J., & Gaitzsch, L. (2019). Adaptive learning system in automotive software engineering. *In 27th international conference on software, telecommunications and computer networks (SoftCOM), IEEE Computer Society, September 2019*. ISBN: 978-1-7281-3711-7.

Tudevdagva, U., & Hardt, W. (2011). *A new evaluation model for e-learning programs*. Chemnitz: Chemnitz University of Technology. Technical Report CSR-11-03.

Tudevdagva, U., & Hardt, W. (2012). A measure theoretical evaluation model for e-learning programs. In *Proceedings of the IADIS on e-Society, Berlin, Germany* (pp. 44–52).

Tudevdagva, U., Hardt, W., Tsoy, E. B., & Grif, M. G. (2012). New approach for e-learning evaluation. In *Proceedings of the 7th international forum on strategic technology 2012, September 17–21, 2012, Tomsk, Russia* (pp. 712–715).

Tudevdagva, U., Hardt, W., & Jargalmaa, D. (2013a). The development of logical structures for e-learning evaluation. In *Proceedings of the IADIS on e-learning 2013, Prague, Czech Republic* (pp 431–435).

Tudevdagva, U., Heller, A., & Hardt, W. (2013b). A model for robustness evaluation of embedded systems. In *Proceedings of the IFOST 2013 conference* (pp. 288–292). Ulaanbaatar

Tudevdagva, U. (2014) Structure oriented evaluation model for e-learning. In *Wissenschaftliche Schriftenreihe Eingebettete Selbstorganisierende Systeme, Universitätsverlag Chemnitz, Chemnitz, Germany, July 2014* (123 p.). ISBN: 978-3-944640-20-4, ISSN: 2196–3932.

Tudevdagva, U., Tomorchodor, L., & Hardt, W. (2014a). The Beta version of implementation tool for SURE model. In *11th joint conference on knowledge-based software engineering (JCKBSE 2014), IEEE Computer Society, September 17–20, 2014, Volgograd, Russia*. In journal of communications in computer and information science, vol. 466 (pp. 243–251).

Tudevdagva, U., Bayar-Erdene, L., & Hardt, W. (2014b). A self-assessment system for faculty based on the evaluation SURE model. In *Proceedings of the 5th international conference on industrial convergence technology, ICICT2014, IEEE computer society, 10–11 May, 2014, Korea* (pp. 266–269). ISBN 978-99973-46-29-2.

Tudevdagva, U., Hardt, W., & Bayar-Erdene, L. (2014c). The SURE model for evaluation of complex processes and tool for implementation. In *The 9th International Forum on Strategic Technology (IFOST 2014), IEEE computer society, October 21–23, 2014*. Bangladesh: Chittagong University of Engineering and Technology.

Tudevdagva, U., Jargalmaa, D., & Bayar-Erdene, L. (2014d). Case study of structure oriented evaluation model. In *Proceedings of the international summer school on e-learning, embedded system and international cooperation, SS2014, 7–13 July, 2014, Laubach, Germany* (pp. 41–44). ISSN 0947-5125.

Tudevdagva, U., Heller, A., & Hardt, W. (2020). An implementation and evaluation report of the active learning method EduScrum in flipped class. In *ICEIT 2020 proceedings, educational and information technology conference, February 11–13, 2020*. Oxford: University of Oxford.

Chapter 4
Online Tool of Data Processing for the SURE Model

4.1 Introduction

This section covers description of online tool, guidance for use and some examples which are intended to assist the use of online tool.

One of the new aspects of SURE model is data processing part. Evaluation models usually do not include an own data processing method. In contrast to that the SURE model includes a data processing part which was developed with help of general measure theory and is strongly linked to the logical structures of evaluation goals. Previously, a test version of online tool for SURE model was used for data collection of some application and was not further developed (Tudevdagva et al. 2014).

The calculation of an evaluation score based on the SURE model involves several steps and can be tedious if the sample size is large. Therefore an online tool has been developed for data processing to support application of the SURE model in evaluation processes.

A corresponding online tool can be found here: http://uranchimeg.com/sure/eva. php

4.2 Data Processing Algorithm

The algorithm of data processing tool is presented in Fig. 4.1.

The algorithm includes the following:

- Input of collected data in CSV format;
- Check of the entered data;
- If data format or structure is not correct an error message is generated;
- Data normalization;

© Springer Nature Switzerland AG 2020
U. Tudevdagva, *Structure-Oriented Evaluation*,
https://doi.org/10.1007/978-3-030-44806-6_4

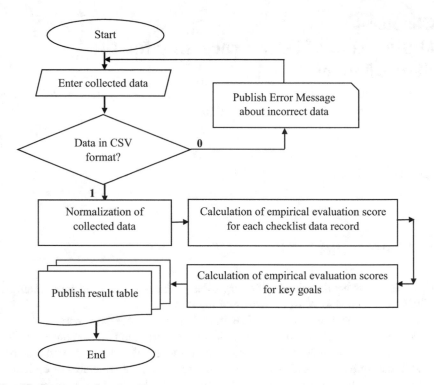

Fig. 4.1 Data processing algorithm

- Calculation of empirical evaluation scores;
- Calculation of empirical evaluation scores for sub goals;
- Creation of evaluation table

Figure 4.2 shows basic algorithm of online tool.

4.3 Properties of Online Tool

To given checklist data the following calculations are carried out:

- Calculation of empirical evaluation scores $Q_e^{*(k)}(C)$ for each checklist data record

$$x_{11}^{(k)}, \ldots, x_{1s_1}^{(k)}, \ldots, x_{r1}^{(k)}, \ldots, x_{rs_r}^{(k)},$$

$k = 1, \ldots, n$, n—sample size.
- Calculation of empirical evaluation score $Q_e^*(C)$ as arithmetic mean over the empirical evaluation scores $Q_e^{*(1)}(C), \ldots, Q_e^{*(n)}(C)$ of checklist data records. In

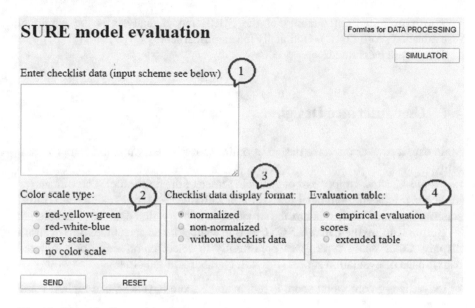

Fig. 4.2 View of online tool for SURE model, upper part

analogous manner the empirical key goal evaluation scores $Q_e^*(B_i), i = 1, \ldots, r$, are calculated, see Sect. 4.2.

- Optionally, the empirical scores $Q^*(C)$ and $Q^*(B_i), i = 1, \ldots, r$, are calculated.
- For sample sizes $n \geq 10$ asymptotic confidence intervals for $Q_e(C)$ are computed.
- An overview of the formulas used can be found in Sect. 4.2 or online here: http://uranchimcg.com/surc/cva_formulas.html
- The results of checklist data analysis are summarized in an evaluation table.
- After entering of evaluation structure parameters and checklist data the calculation begins, the entered data are checked for structural correctness. If yes, a SURE model evaluation summary is created with the relevant empirical evaluation scores. If data do not fit into predefined evaluation structure, a corresponding error message is issued.

The tool has a simulation unit which can be used to generate random checklist data for a given evaluation structure. With the help of this simulator the use of the online tool can be trained and a certain feeling for the SURE model evaluation and the interpretation and understanding of SURE model analyses can be developed.

Several options are available for the presentation of the evaluation results.

- Colored background of the observed values and obtained empirical scores to a given color scale. Different color scales are available for this purpose. This helps to identify trends in the checklist data.
- Normalized or non-normalized display of observation data.

- The most important quantity of the SURE model analysis is the empirical evaluation score $Q_e^*(C)$. Optionally the empirical scores $Q^*(C)$ and $Q^*(B_i)$, $i = 1, \ldots r$, are calculated.

4.4 User Interface Design

Main aspect of user interface design of online tool is its easy use and comprehensibility.

Figure 4.2 shows upper part of the main screen of the online tool. There an empty input box is located (Number 1 in red circle) with title **"Enter checklist data (input scheme see below)"**. In this box the structure and observation data are to be entered.

Before sending the data via **SEND** button evaluator can select the form of result display. **Color scale type** gives opportunity to select color scale for a colored background of evaluation scores (Fig. 4.2, Number 2 in red circle).

- red-yellow-green: worst score is red, average score is yellow, and highest score is green;
- red-white-blue: worst score is red, average score is white, and highest score is blue;
- gray scale: from dark gray to light gray;
- no color scale: no colored background

Moreover the view of collected data in table can be selected. Moreover the format of checklist data of result display can be selected. If evaluator wants to see the original values then it should be checked non-normalized from "Checklist data display format" (Fig. 4.2, Number 3 in red circle).

- normalized: observation data values are transformed to interval from 0 to 1;
- non-normalized: data are displayed in their original form;
- without checklist: data are not included in table.

Evaluation table can be displayed in two different formats (Fig. 4.2, Number 4 in red circle).

- empirical evaluation scores: table includes empirical evaluation scores only;
- extended table (empirical evaluation and standard scores): empirical scores and empirical evaluation scores are displayed.

There are four buttons in the screen.

- **SEND**: start to process data using online tool and show result;
- **RESET**: reset entered data and return to initial state;
- **Formulas for DATA PROCESSING**: calling up a window with all formulas of the SURE model for data processing;
- **SIMULATOR**: calling up a window where a given evaluation structure can be entered and a corresponding random checklist sample can be generated.

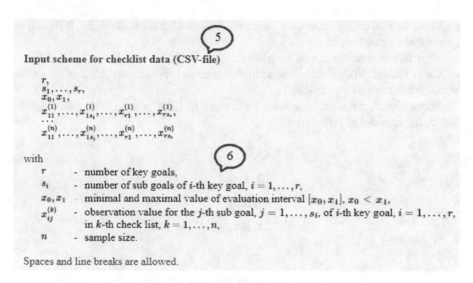

Input scheme for checklist data (CSV-file)

$$
\begin{aligned}
&r, \\
&s_1, \ldots, s_r, \\
&x_0, x_1, \\
&x_{11}^{(1)}, \ldots, x_{1s_1}^{(1)}, \ldots, x_{r1}^{(1)}, \ldots, x_{rs_r}^{(1)}, \\
&\quad \cdots \\
&x_{11}^{(n)}, \ldots, x_{1s_1}^{(n)}, \ldots, x_{r1}^{(n)}, \ldots, x_{rs_r}^{(n)}
\end{aligned}
$$

with

r	-	number of key goals,
s_i	-	number of sub goals of i-th key goal, $i = 1, \ldots, r$,
x_0, x_1	-	minimal and maximal value of evaluation interval $[x_0, x_1]$, $x_0 < x_1$,
$x_{ij}^{(k)}$	-	observation value for the j-th sub goal, $j = 1, \ldots, s_i$, of i-th key goal, $i = 1, \ldots, r$, in k-th check list, $k = 1, \ldots, n$,
n	-	sample size.

Spaces and line breaks are allowed.

Fig. 4.3 View of online tool for SURE model, lower part

The lower part of basic screen shows format of input data with their explanation (Fig. 4.3).

Data have to be entered as comma separated variables (CSV). An example of the structure of the input data is described on screen (Fig. 4.3, Number 6 in red circle). The CSV file contains comma separated the following data:

- r—number of key goals;
- $s_1, \ldots, s_i$—numbers of sub goals;
- x_i, x_j—evaluation interval;
- $x_{11}^1, \ldots, x_{rsr}^n$—checklist data records;

In general, user design of online tool is very easy. Not so many colors and buttons.

4.5   Online Simulator for the SURE Model

Calculation of scores by SURE model is a bit complex. To understand details of data processing and to get a certain feeling for the model simulated data based on a given goal structure can be used. For this purpose an online simulator for the SURE model was integrated. This tool can be reached via the button SIMULATOR of main screen or here: http://uranchimeg.com/sure/eva_simul.html

Assume the following goal structure is given for evaluation: five key goals, the first key goal consists of two sub goals, the second key goal has single sub goal, the third key goal consists of three sub goals, the fourth key goal has two sub goals, and the last fifth key goal has only single sub goal. The evaluation interval goes from 1 to 5 and the evaluation values are randomly selected from set $\{1,\ldots,5\}$. It should be generated a sample of size $n = 5$.

How this evaluation can be simulated.

Call "SIMULATOR" from online tool. Figure 4.4 shows screenshot of the online simulator for the SURE model.

When simulator window comes enter data in CSV format. For example, the following data are to be entered:

$$5,$$

$$2, 1, 3, 2, 1,$$

$$1, 5,$$

$$1, 5,$$

$$5.$$

Figure 4.5 shows screen with entered data in CSV format. To generate random data, press the button **SEND**.

Online tool returns generated data in new screen (Fig. 4.6).

This screen contains a considered evaluation structure as well as a corresponding simulation data set.

To process generated data by online tool one has to press the button "SURE model evaluation."

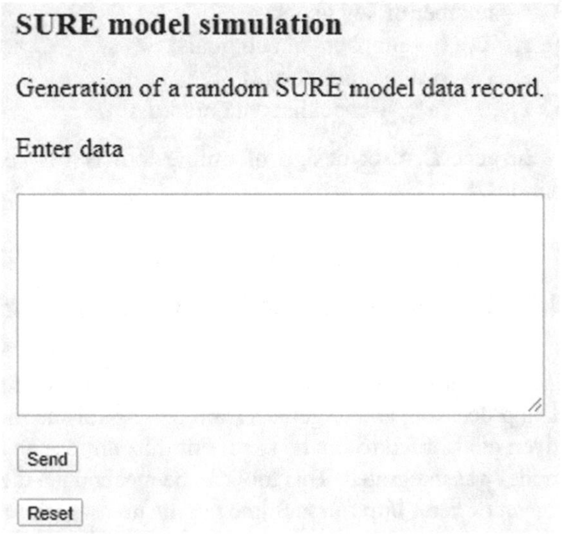

Fig. 4.4 SURE model simulation

SURE model simulation

Generation of a random SURE model data record.

Enter data

Send

Reset

Fig. 4.5 Entered input in
CSV format to simulator

SURE model simulation

Generation of a random SURE model data record.

Enter data

```
5,
2,1,3,2,1,
1,5,
1,5,
5
```

Fig. 4.6 Random data
simulation

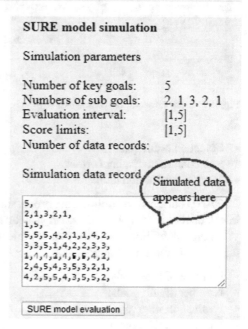

SURE model simulation

Simulation parameters

Number of key goals: 5
Numbers of sub goals: 2, 1, 3, 2, 1
Evaluation interval: [1,5]
Score limits: [1,5]
Number of data records:

Simulation data record   Simulated data
                          appears here

```
5,
2,1,3,2,1,
1,5,
5,5,5,4,2,1,1,4,2,
3,3,5,1,4,2,2,3,3,
1,1,1,1,1,5,5,4,2,
2,4,5,4,3,5,3,2,1,
4,2,5,5,4,3,5,5,2,
```

SURE model evaluation

4.6   Online Data Processing

If data are collected and prepared in correct format, call online tool via
http://uranchimeg.com/sure/eva.php
Enter data into empty input box in CSV format and click on the button **SEND**.
Screen which is shown in Fig. 4.7 is basic screen of the online SURE model.
There are 3 options for the design of the result display: Selection of the "Color scale
type" for color highlighting of checklist data and empirical scores, selection of the
"Checklist data display" format (normalized, non-normalized or without checklist
data) and selection of the "Evaluation table" type (only empirical evaluation scores,
extended table with empirical and empirical evaluation scores).

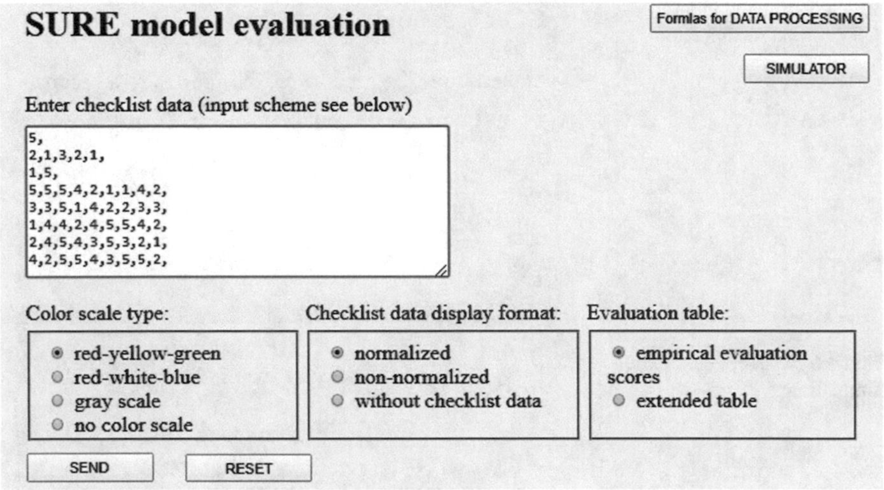

SURE model evaluation   Formlas for DATA PROCESSING

                                                     SIMULATOR

Enter checklist data (input scheme see below)

```
5,
2,1,3,2,1,
1,5,
5,5,5,4,2,1,1,4,2,
3,3,5,1,4,2,2,3,3,
1,4,4,2,4,5,5,4,2,
2,4,5,4,3,5,3,2,1,
4,2,5,5,4,3,5,5,2,
```

Color scale type: Checklist data display format: Evaluation table:

  ● red-yellow-green   ● normalized   ● empirical evaluation
  ○ red-white-blue   ○ non-normalized   scores
  ○ gray scale   ○ without checklist data   ○ extended table
  ○ no color scale

  SEND   RESET

Fig. 4.7 Basic screen of the online tool with input data)

- Selection color scale;
- Selection data display format;
- View of evaluation table

SEND button starts online data processing. This action returns calculated scores in selected type of table and logical structure of evaluation goals (Fig. 4.8).

In Fig. 4.8 is shown only the part of table with empirical evaluation scores. Beside this table on the screen appears:

- Number of key goals;
- Number of sub goals;
- Evaluation interval;
- Number of data records;
- Evaluation structure;
- Evaluation table;
- Color scale in table;
- Empirical evaluation score;
- Asymptotic confidence intervals;
- Empirical key goal evaluation scores;
- Single empirical evaluation scores;
- Button **SURE model formulas**;
- Evaluation data set

These data give transparent overview about evaluation. From this screen one can get complete information on evaluation process. Moreover one can see how many key and sub goals were defined, which evaluation interval is used, what is the sample size for data processing. Table with evaluation scores can be in color or without

Evaluation table (empirical evaluation scores)

k	B_1		B_2	B_3			B_4		B_5	$Q^*_{e,k}(C)$
	A_{11}	A_{12}	A_{21}	A_{31}	A_{32}	A_{33}	A_{41}	A_{42}	A_{51}	
1	5	5	5	4	2	1	1	4	2	0.56
2	3	3	5	1	4	2	2	3	3	0.53
3	1	4	4	2	4	5	5	4	2	0.62
4	2	4	5	4	3	5	3	2	1	0
5	4	2	5	5	4	3	5	5	2	0.68
$Q^*(A_{ij})$	0.5	0.65	0.95	0.55	0.6	0.55	0.55	0.65	0.25	$Q^*_e(C) = 0.477$
$Q^*_c(B_i)$	0.63		0.95	0.77			0.66		0.25	

Evaluation interval $[x_0, x_1] = [1, 5]$

Color scale

1	1.4	1.8	2.2	2.6	3	3.4	3.8	4.2	4.6	5
0	0.1	0.2	0.3	0.4	0.5	0.6	0.7	0.8	0.9	1

Empirical evaluation score $Q^*_e(C)$

$$Q^*_e(C) = 0.477026$$

Fig. 4.8 Processed data in selected table

color. Colored tables facilitate the interpretation of the evaluation result. They also help the evaluator to recognize tendencies in the observation data.

4.7 Formulas for Data Processing

The online tool of the SURE model includes a button which calls a special window which shows all necessary formulas for data processing (Fig. 4.9).

The following formulas are listed:

- Goal structure;
- Theoretical scores: score of C;
- Theoretical scores: evaluation scores of C;
- Empirical scores: for k-th checklist data record;
- Empirical scores: for total sample;
- Empirical evaluation scores: for k-th checklist data record;
- Empirical evaluation scores: for total sample

The theoretical background of formulas is explained in detail in Sect. 4.2.

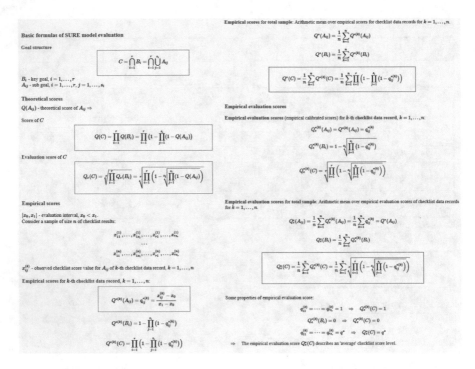

Fig. 4.9 Formulas of data processing

4.8   Summary

This chapter describes the use of online tool for SURE model. The first subsection covers data processing algorithm. The algorithm shows how online tool processes entered data. The second subsection describes basic properties of online tool. Next subsection explains user interface design of the online tool. The fourth subsection covers online simulator for SURE model. The fifth subsection describes online data processing.

Main purpose of this online tool is to support evaluators and researchers for their usage of SURE model. This online tool helps the evaluators to reduce time for data processing and provides a detailed and structured overview on obtained evaluation results. Moreover online tool includes a survey on formal background of calculated scores.

References

Tudevdagva, U., Tomorchodor, L., & Hardt, W. (2014). The beta version of implementation tool for SURE model. In *11th joint conference on knowledge-based software engineering (JCKBSE 2014), IEEE computer society, September 17–20, 2014, Volgograd, Russia*. In Journal of Communications in Computer and Information Science, vol. 466 (pp. 243–251). ISSN: 2196-3932.

Chapter 5
Application of SURE Model and Its Future Development

The structure oriented evaluation model was developed originally for evaluation of e-learning courses. But the idea to use logical structures for definition of evaluation goals opens an opportunity to broaden the field of applications and sciences of this model.

Since 2011 author started to distribute this idea via her works in international conferences (Tudevdagva and Hardt 2011, 2012; Tudevdagva et al. 2013a, 2013b, 2014; Tudevdagva and Bayar-Erdene 2015, 2016; Bayar-Erdene and Tudevdagva 2018). Latest one is a paper which is accepted for 2020 9th International Conference on Educational and Information Technology (ICEIT 2020) which held in February 11–13, 2020 at St Anne's College, University of Oxford, United Kingdom. This paper invited to publish in the International Journal of Information and Education Technology (IJIET: http://www.ijiet.org), which will be indexed by Scopus (Since 2019), EI (INSPEC, IET), EBSCO, Electronic Journals Library, Google Scholar, Crossref, etc.

The presentation and discussion of SURE model in different communities, scientific groups, and conferences has confirmed that the idea of using logical structures to define evaluation goals and the use of data processing rules which are adapted to the logical goal structure is accepted. The theoretical background of the data processing part of SURE model recommends this model as a versatile evaluation tool for evaluation and control of complex processes and systems such as robotics, for instance.

In 2013 A. Heller used structure oriented evaluation model for her doctoral thesis for evaluation of robustness of robots (Heller 2013).

In March of 2019 L. Bayar-Erdene applied structure oriented evaluation model for his doctoral thesis as base evaluation method and model for performance assessment of faculty members (Bayar-Erdene 2019).

The structure oriented evaluation model has space for future development in theory and application. Beside of expanding the areas of application for logically oriented evaluation in complex systems and processes, this can include taking into

© Springer Nature Switzerland AG 2020
U. Tudevdagva, *Structure-Oriented Evaluation*,
https://doi.org/10.1007/978-3-030-44806-6_5

account further logical target structures such as alternative target structures or the inclusion of conditional evaluation scores, for instance.

References

Bayar-Erdene, L. (2019). *Evaluation of faculty members by structure oriented evaluation*. Doctoral Thesis, Mongolian University of Science and Technology, Ulaanbaatar.

Bayar-Erdene, L., & Tudevdagva, U. (2018, July) Online self-assessment system based on structure oriented evaluation model. *International Journal of Science and Research (IJSR)*, 7(7), 795–798. ISSN: 2319-7064.

Heller, A. (2013). *Systemeigenschaft Robustheit - Ein Ansatz zur Bewertung und Maximierung von Robustheit eingebetteter Systeme*. PhD Thesis, Win. Schriftenreihe 'Eingebettete, Selbstorgan-isierende Systeme' (Vol. 12). Universitätsverlag Chemnitz.

Tudevdagva, U., & Bayar-Erdene, L. (2015, April). The self-assessment system for faculty, in *Proceedings of Open scientific conference at Plekhanov University, April 2015, Ulaanbaatar*, pp.127–129.

Tudevdagva, U., & Bayar-Erdene, L. (2016). Application of the structure oriented evaluation model fornfaculty members self-assessment. In *Proceedings of 11th International Forum on Strategic Technology IEEE conference, IFOST 2016, May 29-June 2, Novosibirsk, Russia* (pp. 448–451). https://doi.org/10.1109/IFOST.2016.7884292

Tudevdagva, U., & Hardt, W. (2011). *A new evaluation model for e-learning programs*. Technical Report CSR-11-03, Chemnitz University of Technology.

Tudevdagva, U., & Hardt,W. (2012). A measure theoretical evaluation model for e-learning programs. In *Proceedings of the IADIS on e-Society, Berlin* (pp. 44–52).

Tudevdagva, U., Hardt,W., & Jargalmaa, D. (2013a). The development of logical structures for e-learning evaluation. In *Proceedings of the IADIS on e-Learning 2013*, Prag (pp. 431–435).

Tudevdagva, U., Heller, A., & Hardt, W. (2013b). A model for robustness evaluation of embedded systems. In *Proceedings of the IFOST 2013 Conference, Ulaanbaatar* (pp. 288–292).

Tudevdagva, U., Hardt, W., & Bayar-Erdene, L. (2014). The SURE model for evaluation of complex processes and tool for implementation. In *The 9th International Forum on Strategic Technology (IFOST 2014), Chittagong University of Engineering and Technology, Chittagong, October 21–23, 2014*. Washington, DC: IEEE Computer Society.

Index

Printed in the United States
by Baker & Taylor Publisher Services